This series aims to report new developments in physical research and teaching – quickly, informally, and at a high level. The type of material considered for publication includes:

1. Preliminary drafts of original papers and monographs

2. Lectures on a new field, or presenting a new angle on a classical field

3. collections of seminar papers

4. Reports of meetings

Texts which are out of print but still in demand may also be considered if they fall within these categories.

The timeliness of a manuscript is more important than its form, which may be unfinished or tentative. Thus, in some instances, proofs may be merely outlined and results presented which have been or will later be published elsewhere.

Publication of *Lecture Notes* is intended as a service to the international physical community, in that a commercial publisher, Springer-Verlag, can offer a wider distribution to documents which would otherwise have a restricted readership. Once published and copyrighted, they can be documented in the scientific libraries.

Manuscripts

Manuscripts are reproduced by a photographic process; they must therefore be typed with extreme care. Symbols not on the typewriter should be inserted by hand in indelible black ink. Corrections to the typescript should be made by sticking the amended text over the old one, or by obliterating errors with white correcting fluid. The figures (in the original size) ready for reproduction should be inserted into the text. Should the text, or any part of it, have to be retyped, the author will be reimbursed upon publication of the volume. Authors receive 50 free copies.

The typescript is reduced slightly in size during reproduction, therefore a large size of type should be used; best results will not be obtained unless the text on any one page is kept within the overall limit of 18 x 26.5 cm (7 x 10½ inches). **The publishers will be pleased to supply on request special stationery with the typing area outlined.**

Manuscripts in English, German or French should be sent to Springer-Verlag, 6900 Heidelberg, Postfach 1780.

Die *„Lecture Notes"* sollen rasch und informell, aber auf hohem Niveau, über neue Entwicklungen in der Physik berichten. Zur Veröffentlichung kommen:

1. Vorläufige Fassungen von Originalarbeiten und Monographien.

2. Spezielle Vorlesungen über ein neues Gebiet oder ein klassisches Gebiet in neuer Betrachtungsweise.

3. Seminarausarbeitungen.

4. Vorträge von Tagungen.

Ferner kommen auch ältere vergriffene spezielle Vorlesungen, Seminare und Berichte in Frage, wenn nach ihnen eine anhaltende Nachfrage besteht.

Die Beiträge dürfen im Interesse einer größeren Aktualität durchaus den Charakter des Unfertigen und Vorläufigen haben. Sie brauchen Beweise unter Umständen nur zu skizzieren und dürfen auch Ergebnisse enthalten, die in ähnlicher Form schon erschienen sind oder später erscheinen sollen.

Die Herausgabe der *„Lecture Notes"* Serie durch den Springer-Verlag stellt eine Dienstleistung an die physikalischen Institute dar, indem der Springer-Verlag für ausreichende Lagerhaltung sorgt und einen großen internationalen Kreis von Interessenten erfassen kann. Durch Anzeigen in Fachzeitschriften, Aufnahme in Kataloge und durch Anmeldung zum Copyright sowie durch die Versendung von Besprechungsexemplaren wird eine lückenlose Dokumentation in den wissenschaftlichen Bibliotheken ermöglicht.

Lecture Notes in Physics

Edited by J. Ehlers, München, K. Hepp, Zürich and
H. A. Weidenmüller, Heidelberg
Managing Editor: W. Beiglböck, Heidelberg

10

John M. Stewart

Institute of Theoretical Astronomy,
Cambridge, England

Non-Equilibrium Relativistic Kinetic Theory

Springer-Verlag Berlin Heidelberg GmbH 1971

ISBN 978-3-540-05652-2 ISBN 978-3-540-36940-0 (eBook)
DOI 10.1007/978-3-540-36940-0

 Library of Congress Catalog Card Number 70-179436

Originally published by Springer-Verlag Berlin Heidelberg New York in 1971.

CONTENTS

CHAPTER 1. INTRODUCTION

In the last few years a number of papers on relativistic statistical mechanics and kinetic theory have been published. There are many reasons for this new interest. In cosmology Misner (1968) and Stewart (1969) have suggested that non-equilibrium processes are of fundamental importance in the early stages of the universe. Interest in statistical mechanics has been awakened by the recent theories of the early universe due to Hagedorn (1970), and Omnes (1970). The study of quasars has aroused interest in relativistic stellar clusters, (for a recent review see eg. Fackerell et al (1969)) and in astrophysics the theory is relevant whenever gravitational fields are large, (eg in neutron stars,) or plasmas are relativistic,(eg sources of sychrotron emission,) or in degenerate electron gases whose Fermi energy is of the order of $m_e c^2$, as in some white dwarfs. Fairly recent reviews of relativistic stellar structure have been given by Thorne (1967, 1971) and some of the problems raised by Thorne which could be tackled include the study of non-adiabatic perturbations to a gas in thermal equilibrium, and the relativistic energy transport theory for stellar interiors. Other good reasons for studying relativistic statistical mechanics and kinetic theory are that the inherent simplicity of the spacetime picture may throw new light on the well known results of the classical theory, and that it may be possible to answer the basic question as to the compatibility of relativistic and statistical mechanics.

The formulation of statistical mechanics within the framework of relativity theory is still in a controversial stage. A clear but out of date review has been given by Havas (1965) and so we shall summarise some of the more recent developments. A first group of authors including Havas (1965), Bergmann (1951), Bakamjian & Thomas (1953), Foldy (1961) and Krizan (1965) hold that relativistic statistical mechanics should be formulated in terms of the coordinates of a finite number of particles alone. However this formalism cannot give rise to a general Hamiltonian theory of relativistic dynamics. Currie (1963) has proved that such a formalism is compatible with the concept of invariant trajectories of material particles only for free particles. Van Dam & Wigner (1965) have attempted to get round this difficulty, but their formulation is definitely non-Hamiltonian.

Another point of view consists of treating relativistic systems as a collection of particles interacting through a field. This idea has been developed by a number of authors including Prigogine (1965), Mangeney (1964,1970), De Gottal & Prigogine (1965), Klimontovich (1958,1960), Balescu & Kotera (1967). The theorem of Currie is then no longer applicable, and invariant trajectories are described by the theory. The theory is very rich; it describes not only fluid mechanics but also the evolution of the field, including all possible radiation, absorbtion and scattering processes. The main difficulty is that the infinite number of degrees of freedom leads to the well known divergence problems of field theory.

Authors of a third group including Bergmann (1951), Van Dantzig (1939), Hakim (1968), Gudehus & Dubin (1969), Müller (1970) have tried a more general approach. They describe the particles by a set of trajectories in spacetime; this leads to an 8N-dimensional phase space, where N is the number of particles. Philosophically speaking this gives rise to a "static" description of the world, in which past, present and future are given once and for all, and the idea of evolution, and hence of dynamics, is somewhat irrelevant. If $N = 1$ this is not a drawback because one can easily identify the fourth component x^0 of the particle event with the observer's time. But when $N > 1$ this identification is no longer possible. The various fourth components $x^0_{(i)}$ of the particles must be interrelated. This relation does not follow from dynamics, but is arbitrary. This has been formally demonstrated by Dirac (1950, 1951) who has shown that a Hamiltonian theory constructed along these lines necessarily contains a large number (N) of arbitrary functions which can be eliminated only by imposing (arbitrary)relations between the individual particles' proper times and the observer's time. The situation is even worse if the field is added.

The case N = 1 corresponds to kinetic theory and seems to be the most satisfactory theory to study at the present time although it is somewhat limited in application. Recent comprehensive reviews of the subject have been given by Sachs (1971), Ehlers (1971), and earlier papers include Bergmann (1951), Synge (1957), Sasaki (1958), Tauber & Weinberg (1961), Ehlers (1961), Chernikov (1963), Israel (1963), Bichteler (1965,1967), Lindquist (1966), Marle (1969), Stewart (1969), Anderson (1970), and Stewart & Anderson (1971). The 1-particle phase space is the space which contains the spaces of all 4-momentum vectors at each point of spacetime. This space is called the tangent bundle over spacetime, and kinetic theory can be formulated in a geometrically appropriate way using the language of fibre bundle theory. This approach has been emphasised by Lindquist (1966), and Ehlers (1971). However since many physicists will not be acquainted with these mathematical notions, we give a brief heuristic summary of the foundations in section 2.1, while the more formal, rigorous approach is summarised in sections 2.2 - 2.4. In section 2.5 the Boltzmann equation is discussed, and section 2.6 derives the H-theorem and the equilibrium distributions. It is shown there that equilibrium distributions fall into two classes,

i) the collision free gas,

ii) the collision-dominated gas in which "detailed balancing" occurs.

Distributions in the second category have a Maxwellian or Bose-Einstein (Fermi-Dirac distribution function. Further it is shown that the only motion which such a gas can have is rigid, (ie. non-expanding, non-shearing, possibly rotating). This theorem provides further motivation for studying non-equilibrium distributions, and the rest of the work is concerned with them.

Before embarking on a detailed study of non-equilibrium kinetic theory chapter 3 discusses relativistic thermodynamics. In the classical theory the thermal properties of a body are divorced from its motion, and there is therefore no indication of how to generalise the theory to the relativistic case. Because of this a large number of different formulations have been developed, and in particular there has been a heated discussion of the Lorentz transformation properties of the temperature; does a moving body appear cool or not? (See eg. Arzelies (1965), Møller (1967), Landsberg & Johns (1970), and references cited there.) However it is shown that there is essentially only one theory of relativistic thermodynamics which is compatible with the kinetic theory developed in chapter 2. This was developed independently by Kluitenberg & de Groot (1953) and Stueckelberg & Wanders (1953). A formulation of this theory in modern notation is given in chapter 3. In particular we develop by a plausibility argument the phenomenological linear transport theory for small deviations from equilibrium.

In chapter 4 we discuss approximate non-equilibrium solutions of the Boltzmann equation. Previous studies have been based either on the Chapman-Enskog method, (Sasaki (1958), Israel (1963), Marle (1969),) or the Grad method of moments, (Chernikov (1963), Marle (1969), Stewart (1969), Anderson (1970)). A discussion of these approaches is given in section 4.1. Israel was the first author to explicitly demonstrate the existence of a bulk viscosity for a relativistic gas. This result was later confirmed by Marle, Stewart and Anderson, and supplies the physical reason why, under certain circumstances, an expanding relativistic gas cannot be in equilibrium. The work of Marle (1969) is independent of, but very similar to, and mathematically more complete than, the work of Stewart & Anderson. Marle considers a 1-component gas with a simplified model for the collision term in the Boltzmann equation. The work of Stewart (1969) and Anderson (1970) is rather more general since the full Boltzmann collision term is studied. In chapter 4 the generality is increased further. There we study a multicomponent gas with chemical reactions, annihilation and creation processes included as well. In this respect the relativistic treatment presented here is slightly more general than the classical one, (see eg. Burgers (1969)), and this gain is due to the fact that the Grad method is actually simpler for the relativistic case than for the classical one, for reasons which are explained in chapter 4. We develop the linear transport theory and derive formulae for the coefficients of heat conduction, bulk and shear viscosity, diffusion, and reaction rates.

A major flaw in all previous treatments of relativistic thermodynamics is that the heat conduction equation leads, just as in the classical case, to a parabolic equation for the propagation of the temperature, and this (theoretically) permits an infinite velocity for heat signals. This clearly violates relativity theory. In section 4.7 we consider arbitrary non-adiabatic perturbations from thermal equilibrium, and show that the maximum propagation (phase) velocity for such disturbances is $c\sqrt{3/5} \approx 0.8c$.

The collision integrals required in the Grad theory are usually 9 or 12-fold, and in chapter 5 we show how to reduce these to 2-fold integrals over the relative momentum and scattering angle. This final integration can be accomplished once the scattering cross-section is specified, and as an example we evaluate the transport coefficients explicitly for a 1-component gas with constant cross-section.

We have deliberately refrained from giving more realistic examples. The object of this study is to provide a tool with uses in many branches of relativistic physics. Each application would constitute a research project in itself.

The sign convention and choice of units used here follows the spacelike convention proposed by Misner et al (1968). The signature is (-+++), and if u^a is a timelike vector, $u^a u_a < 0$. Although the notation is general relativistic, the theory can easily be restricted to special relativity. To effect this specialisation one should replace all covariant (semi-colon) derivatives by partial (comma) derivatives, eg. $u_{a;b} \longrightarrow u_{a,b} = \partial u_a / \partial x^b$, and always take the metric tensor g_{ab} to be diag (-1,1,1,1), the Minkowski tensor. The usual conventions are adopted for symmetrisation etc. of indices, eg.

$$A_{(ab)} := (A_{ab} + A_{ba})/2! ,$$

$$A_{[ba]} := (A_{ab} - A_{ba})/2! .$$

The notation := is taken over from the ALGOL computer language to signify a definition of the quantity occurring on the left hand side of the equation. Greek indices range over 1,2,3.

I would like to thank Sidney Sussex College Cambridge and the Institute of Theoretical Astronomy Cambridge for Fellowships while this work was carried out. I would also like to thank Drs. DW Sciama, and GFR Ellis for acting as my supervisors, Professor JL Anderson for collaborating on an earlier version of the theory (Anderson (1970)), Professor AH Taub, Drs W Kundt, SW Hawking, WC Saslaw and Messrs MAH MacCallum and R Treciokas for stimulating discussions. Parts of chapters 2, 5 are based on Ehlers (1971), and I should like to thank Professor Ehlers for his constructive comments following a detailed reading of the manuscript. The calculations leading to the asymptotic series in chapter 5 were performed using the CAMAL algebraic manipulation language on the TITAN computer, Mathematical Laboratory, University of Cambridge.

CHAPTER 2. FOUNDATIONS OF RELATIVISTIC KINETIC THEORY

2.1 Introduction.

As was emphasised in the introduction, the phase space for a system of relativistic particles is simply the tangent bundle over the spacetime manifold, and the corresponding distribution function can naturally be described as a scalar field on this bundle. This leads to an elegant treatment of the theory which will be described in this chapter. However some readers would prefer to see the theory presented at a less sophisticated but more intuitive level, and such a treatment will be given in this section. This chapter can therefore be read in one of three ways:

i) the physicist not acquainted with modern differential geometry can read this section, and then jump to section 2.5,

ii) those already acquainted with relativistic kinetic theory can skip the rest of this section, and start at section 2.2,

iii) the general reader can read through the chapter in the normal way. In this case this section serves as an introduction to the abstract concepts presented in the next three sections.

Let X be the spacetime manifold, and let x be a point of X. The tangent vectors at x span a vector space, the <u>tangent space</u> T_x at x. By a suitable parametrisation of the particle worldlines we can identify tangent vectors with 4-momenta, and so T_x becomes the natural relativistic generalisation of <u>momentum space</u>. (Strictly speaking, physically realisable 4-momenta are timelike or null. This point will be considered later.) We now want to define a distribution function $f(x,p)$ which describes the number dN of particles which cross a certain spacelike volume element dV at x, and whose 4-momenta lie within a certain volume element dP in momentum space. It is clearly desirable to choose dV, dP in a coordinate (Lorentz-) invariant way; then since dN is invariant $f(x,p)$ will be Lorentz-invariant.

We start by defining dV. Let u^a be an arbitrary 4-velocity, ($u_a u^a = -1$, $u^0 > 0$,) at a given point x, and let d_1x^a, d_2x^b, d_3x^c be three arbitrary displacement vectors at x which span an element of hypersurface orthogonal to x. This condition can be written,

$$\sigma_a := \eta_{abcd}\, d_1x^b d_2x^c d_3x^d = \text{const.} \times u_a \,,$$

where η_{abcd} is the completely antisymmetric 4-tensor with $\eta_{0123} = \sqrt{-g}$, and $g = \det(g_{ab})$. If we call the constant of proportionality $-dV$ then,

$$dV = \eta_{abcd}\, u^a d_1x^b d_2x^c d_3x^d \,,$$

and

$$\sigma_a = -dV\, u_a .$$

dV or equivalently σ_a is our required volume element. For an observer with 4-velocity u^a who uses orthonormal coordinates, $dV = d^3x$ as usual. Clearly a coordinate independent volume element on momentum space is,

$$dP := \eta_{abcd}\, d_0p^a d_1p^b d_2p^c d_3p^d \,,$$

where d_Ap^a, $(A=0,1,2,3)$ are 4 arbitrary displacement vectors in momentum space. For the moment we shall consider a gas in which all the particles have rest mass m and so the 4-momenta must satisfy,

a) $p^0 > 0$,

b) $p_a p^a = -m^2$.

Condition b) is a limitation on the parts of momentum space which are physically accessible, and clearly a coordinate independent volume element for this part of phase space is

$$\pi_m := 2H(p)\,\delta(p_a p^a + m^2)\, dP,$$

where,

$$H(p) := \begin{cases} 1 & \text{if } p^0 > 0, \\ 0 & \text{otherwise.} \end{cases}$$

Only three components of the 4-momentum are independent and so we may only construct three arbitrary momentum displacement vectors d_1p^a, d_2p^b, d_3p^c. If we take p^1, p^2, p^3 as independent components the usual rules for dealing with δ functions give, taking condition a) into account,

$$\pi_m = \sqrt{-g}\,|p_0|^{-1}\,\varepsilon_{\alpha\beta\gamma}\,d_1p^\alpha d_2p^\beta d_3p^\gamma . \qquad (2.11)$$

In locally orthonormal coordinates at x, $\pi_m \longrightarrow d^3p/E$ where d^3p is the usual momentum space volume element, and E is the energy of a particle.

In ordinary kinetic theory we represent an instantaneous gas state by a cloud of points in a 6-dimensional phase space, each point corresponding to a particle. The distribution function is the number density of the cloud, (more exactly an ensemble average,) and its evolution determines the behaviour of the gas. However in a relativistic spacetime there are no preferred time slices $t =$ constant, and hence it is appropriate to take the phase space to be seven dimensional, with coordinates (x^a, p^α). The picture of the gas is then a static one, consisting of the world lines of the particles, (see fig.1,) and accordingly we should define f in terms of the number dN of world lines which intersect a spacelike hypersurface in X with momenta in a given range. Between collisions the worldlines satisfy the equations,

$$\frac{dx^a}{dv} = p^a \quad , \quad \frac{dp^a}{dv} = -\Gamma^a{}_{bc}\,p^b p^c , \qquad (2.1)$$

where v is an affine parameter, and electromagnetic forces have not been taken into account. Thus the worldlines are not directed normally to the spacelike volume element, and so we must include a projection factor $(-u_a p^a)$ in our definition of $f(x,p)$,

$$dN = f(x,p)(-u_a p^a)\,dV\,\pi_m .$$

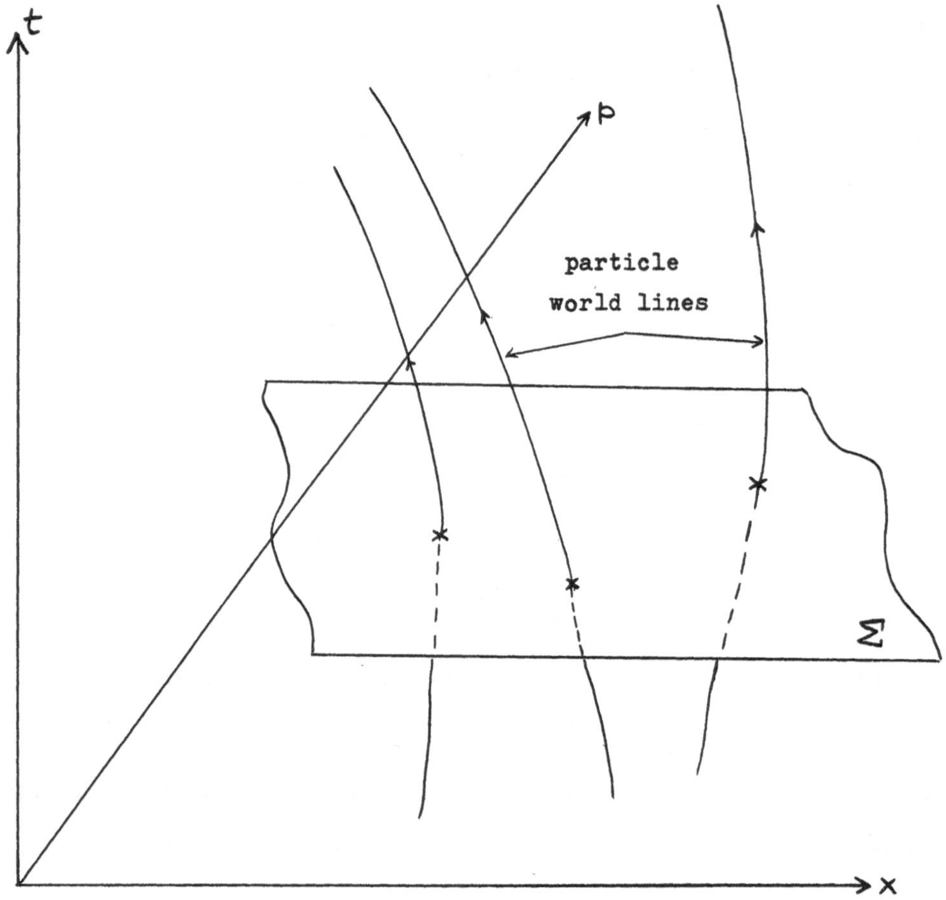

Figure 1. A 3-dimensional representation of the 7-dimensional phase space. Each worldline represents the complete history of a particle. Each spacelike hypersurface (slice) Σ, (eg.t = const.) is cut by the worldlines in a set of points. These slices correspond to the classical 6-dimensional phase space in which the instantaneous state of the gas is represented by a set of points which move as the gas evolves.

It is easily verified that for an observer using local orthonormal coordinates, $dN = f(x,p)d^3xd^3p$, and so $f(x,p)$ has its usual meaning for all local observers

Just as in the classical theory we have a Liouville theorem which states that for an arbitrary observer with 4-velocity u^a,

$$(-u_a p^a) dV \pi_m \qquad \text{is conserved along the worldlines.}$$

It follows that the rate of change of dN is,

$$\frac{d(dN)}{dv} = \frac{df}{dv}(-u_a p^a) dV \pi_m,$$

$$= \left(\frac{df}{dx^a}\frac{dx^a}{dv} + \frac{df}{dp^a}\frac{dp^a}{dv}\right)(-u_a p^a) dV \pi_m$$

$$= L(f)(-u_a p^a) dV \pi_m,$$

where L is the operator,

$$L := p^a \frac{\partial}{\partial x^a} - \Gamma^a{}_{bc} p^b p^c \frac{\partial}{\partial p^a} .$$

Two queries can be raised at this stage:

a) is L a coordinate independent operator?

b) since only three of the momentum componants are independent why does L treat all four as independent quantities?

a) is simply answered by noting that L is the directed derivative along the worldlines, and this property is clearly coordinate independent. However the second query is more subtle. The particles are constrained to move on the hypersurfaces $m =$ constant, where,

$$m^2 := -g_{ab} p^a p^b .$$

It is easily verified that $L(m) = 0$, that is, a particle cannot change its rest mass. This means that L considered as a directed derivative acts tangentially on the surfaces $m =$ constant in phase space, and so no difficulty arises because of the non-independence of the momentum components.

If we denote the change in dN due to interactions by

$$C[f]\ (-u_a p^a)\, dV\, \pi_m$$

we obtain the relativistic generalisation of Boltzmann's equation,

$$L(f) = C[f] \quad . \tag{2.17}$$

A lemma which is frequently used in the theory can be stated as follows:

Let

$$T^{ab\ldots cd}(x) := \int p^a p^b \ldots p^c p^d\, G(f)\, \pi_m \ ,$$

where G is an arbitrary function of f and the integral is over all of the surface $m =$ constant in phase space. Then,

$$T^{ab\ldots cd}{}_{;d}(x) = \int p^a p^b \ldots p^c\, G'(f)\, L(f)\, \pi_m ,$$

as can be verified by the methods of section 2.6

In subsequent sections we shall often consider a gas containing particles of differing rest masses. If the A-species has particles of rest mass m_A its momentum space volume element will be written π_A, its distribution function f_A and its Liouville operator L_A.

2.2 Mathematical Preliminaries: Liouville Theorem.

In setting up a relativistic kinetic theory we shall make the usual assumptions upon which Einstein's gravitational theory is based. That is, we shall assume that spacetime is a 4-dimensional oriented differentiable connected Hausdorff manifold X which carries a Riemannian metric g of signature (-+++), and is time oriented with respect to g. The curvature associated with g is related to the matter contained in spacetime by the Einstein field equations,

$$G^{ab} + \Lambda g^{ab} = T^{ab} .$$

The orientability assumption can in fact be deduced from the observed violation of C-invariance if CPT-invariance is assumed. This assumption is strictly not a necessary one since as we shall show, relativistic phase space is always orientable, which is a necessary condition for integration to be defined on it. A convincing argument for time orientability (the existence of a future pointing timelike vector field,) is the observed violation of CP invariance. We shall need this assumption to introduce the collision term in the Boltzmann equation and without this there would be no H-theorem.

Before studying a gas we consider a single charged particle with mass m ($\geqslant 0$), and charge e which moves under the joint influence of a gravitational field, described by the metric tensor g_{ab} and an electromagnetic field described by the Maxwell field tensor F_{ab}. Its worldline $x^a = x^a(v)$ (v is an affine parameter,) is determined by the Lorentz-Einstein equations of motion,

$$\frac{dx^a}{dv} = p^a , \quad \frac{dp^a}{dv} = - \Gamma^a{}_{bc} p^b p^c + eF^a{}_b p^b , \tag{2.1}$$

where radiation reaction has been neglected. The affine parameter v is uniquely determined except for its origin by the requirement that p^a is the 4-momentum. If $m > 0$ we have the alternative definition v = proper time/m. According to equation (2.1) the instantaneous state of a particle is given by a 4-momentum p^a at an event x^a ; these data determine the motion uniquely. Let T_x be the tangent space at a point x in spacetime X, ie. T_x is the vector space of all 4-vectors at x . We define the set

$$M := \left\{(x,p): x \in X,\ p \in T_x,\ p_a p^a \leq 0, p \text{ future directed} \right\} , \quad (2.2)$$

as the one particle phase space for particles of arbitrary rest masses. M is an 8-dimensional manifold with boundary. If (x^a) are local coordinates in X and (p^a) the corresponding components of vectors, then (x^a, p^a) are local coordinates in M. The boundary ∂M consists of all states belonging to zero rest mass. M is a subset of the tangent bundle $T(X)$ of spacetime,

$$T(X) := \left\{(x,p): x \in X,\ p \in T_x \right\} .$$

The equations of motion define on M a vector field,

$$L := p^a \frac{\partial}{\partial x^a} + (eF^a{}_b p^b - \Gamma^a{}_{bc} p^b p^c) \frac{\partial}{\partial p^a} , \quad (2.3)$$

the Liouville vector or operator. The directed parametrised integral curves $(x^a(v), p^a(v))$ of (2.3) form a congruence in M the phase flow generated by L . Physically the phase flow represents the set of all test particle motions which are possible in the combined gravitational and electromagnetic fields occurring in X.

The rest mass m given by,

$$m^2 := -g_{ab}(x) p^a p^b , \quad m \geq 0, \quad (2.4)$$

is a scalar function on M which is constant on each phase orbit,

$$L(m) = 0, \tag{2.5}$$

as can easily be checked by direct differentiation. The hypersurface of M defined by m = constant is generated by all those phase orbits which belong to the assigned mass value; we use the symbol M_m for it. M_m is the phase space for particles of mass m; its dimension is 7 and we take (x^a, p^α) as coordinates on M_m. Because of (2.5) L is tangent to M_m. The restriction of L. to M_m,

$$L_m := p^a \frac{\partial}{\partial x^a} + (eF^\alpha{}_b p^b - \Gamma^\alpha{}_{bc} p^b p^c) \frac{\partial}{\partial p^\alpha} , \tag{2.6}$$

is the Liouville operator associated with M

We have assumed that X is orientable. All of kinetic theory can be developed without this assumption, because M is orientable even if X is not orientable. This follows immediately from the remark that if $(x^a),(x'^a)$ are two coordinate systems on X and if $(x^a,p^a),(x'^a,p'^a)$ are their extensions to M then,

$$\frac{\partial(x'^a,p'^a)}{\partial(x^b,p^b)} = \frac{\partial(x'^a)}{\partial(x^b)} \cdot \frac{\partial(p'^a)}{\partial(p^b)} = \left[\frac{\partial(x'^a)}{\partial(x^b)}\right]^2 \geqslant 0.$$

M_m is also orientable for it is the boundary of that part of M given by $p_a p^a \geqslant -m^2$. We choose the orientation in M by selecting the (x^a,p^a) as a set of oriented coordinate systems, and take the induced orientation in M_m.

Using an oriented coordinate system in the spacetime manifold X we define the 4-form,

$$\eta := \sqrt{-g}\, dx^{0123}, \ (g := \det(g_{ab}), dx^{ab\ldots c} := dx^a \wedge dx^b \wedge \ldots \wedge dx^c). \tag{2.7}$$

The components of η with respect to the oriented coordinate systems are

$$\eta_{abcd} = \eta_{[abcd]} \quad , \qquad \eta_{0123} = \sqrt{-g} \quad .$$

However η is independent of the coordinate system used to define it and is called the (Riemannian) volume element on X.. We can also define a vectorial hypersurface element σ with components,

$$\sigma_a := 1/6 \, \eta_{abcd} \, dx^{bcd} . \qquad (2.8)$$

If is a vector field on X and D a 4-dimensional compact submanifold with boundary ∂D in X henceforth called a region, then Gauss' theorem in a Riemannian space takes the form, (eg. Synge(1964)),

$$\int_D \eta \, A^a{}_{;a} = \int_{\partial D} \sigma_a A^a \quad . \qquad (2.9)$$

The tangent space T_q of a point $q \in X$ is a semi-Riemannian space with a volume element,

$$\pi := \sqrt{-g} \, dp^{0123} \quad . \qquad (2.10)$$

Here g is to be evaluated with respect to the oriented coordinates (x^a), and the (p^a) are associated coordinates in T_q. The physically important hypersurfaces in T_q are the mass shells. The mass shell $P_m(q)$ consists of all future directed 4-momentum vectors p at q which belong to proper mass m, ie. $p^2 = -m^2$. As coordinates on $P_m(q)$ we shall take (p^ν), p^0 is then determined by (2.4). We choose as a coordinate independent volume element on $P_m(q)$,

$$\pi_m := 2H(p)\delta(p^2+m^2)\sqrt{-g}\, dp^{0123} \quad ,$$

where δ is the Dirac distribution function and,

$$H(p) := \begin{cases} 1 \text{ if } p \text{ is future directed,} \\ 0 \text{ otherwise.} \end{cases}$$

The reason for the normalisation factor 2 will appear later. For $m > 0$ $m\,\pi_m$ is the induced Riemannian volume element of $P_m(q)$ considered as a hypersurface in T_q. We can write π_m explicitly as a 3-form,

$$\pi_m = \frac{\sqrt{-g}}{|p_0|}\, dp^{123} \tag{2.11}$$

A coordinate independent volume element on M_m is,

$$\Omega_m := \eta \wedge \pi_m = \frac{(-g)}{|p_0|}\, dx^{0123} \wedge dp^{123} , \tag{2.12}$$

the exterior product of the spacetime volume element η with the measure π_m on the mass shell P_m. The final volume element that we shall need to consider is one for hypersurfaces in M_m. Bichteler (1965) has obtained such a volume element by contracting Ω_m with the Liouville vector L_m, and so we define a volume element for hypersurfaces in M_m by,

$$\omega_m := L_m \cdot \Omega_m$$

$$= p^a \sigma_a \wedge \pi_m + \tfrac{1}{2} |p_0|^{-1} \eta_{0\lambda\mu\nu}(eF^\lambda{}_b p^b - \Gamma^\lambda{}_{bc} p^b p^c) dp^{\mu\nu} \wedge \eta \; . \tag{2.13}$$

ω_m is uniquely determined (except for a numerical factor) by the requirement that it assigns a non-zero volume to any hypersurface element tangent to L_m, (Ehlers (1971)). This property is obviously necessary for our purposes. For if we consider a tube T of phase orbits intersected by a hypersurface, the intersection being Σ, then $\int_\Sigma \omega_m$ is nonzero if and only if the hypersurface is not parallel to the tube. Furthermore ω_m shares an important property with the Lebesgue measure on a Newtonian phase space: it is invariant with respect to the phase flow. To see this we first verify,

$$d\omega_m = 0,$$

by using inertial coordinates (x^a) at an arbitrary event x in X, so that at x, $\Gamma^a{}_{bc} = 0$, $p^0 = (m^2 + p^\nu p^\nu)^{\frac{1}{2}}$, $\partial p^0/\partial x^0 = 0$, and $p^0 dp^0 = p^\nu dp^\nu$ on M_m. Next we apply Stokes' theorem to the tube T of phase orbits bounded by two cross sections Σ, Σ' and the surface Λ of the tube between them, (see fig.2),

$$0 = \int_T d\omega_m = \int_{\partial T} \omega_m = \int_{\Sigma'} \omega_m - \int_{\Sigma} \omega_m + \int_{\Lambda} \omega_m \,.$$

The last term vanishes since L_m is tangent to Λ, and so ω_m vanishes on this surface. Hence

$$\int_{\Sigma} \omega_m = \int_{\Sigma'} \omega_m \,,$$

which expresses the asserted invariance. This result is often referred to as Liouville's theorem.

2.3 The Relation of the Relativistic to the Non-Relativistic Theory.

It is important when comparing the relativistic and classical treatments of kinetic theory to realise that Newtonian phase space does not correspond to the one particle phase space defined here. The actual correspondence is between the 7-dimensional relativistic phase space M_m and the direct product of the time axis with the ordinary 6-dimensional Newtonian one particle phase space, (see fig. 1). Hence a spacelike hypersurface $G \subset X$ at each point of which a momentum region $K_x \subset P_m(x)$ is prescribed, represents a hypersurface $\Sigma \subset M_m$ which corresponds to a part of an "instantaneous" ordinary phase space. On such a Σ, ω_m reduces to its first part,

$$\omega_m = p^a \sigma_a \wedge \pi_m \quad \text{on } \Sigma \,. \tag{2.14}$$

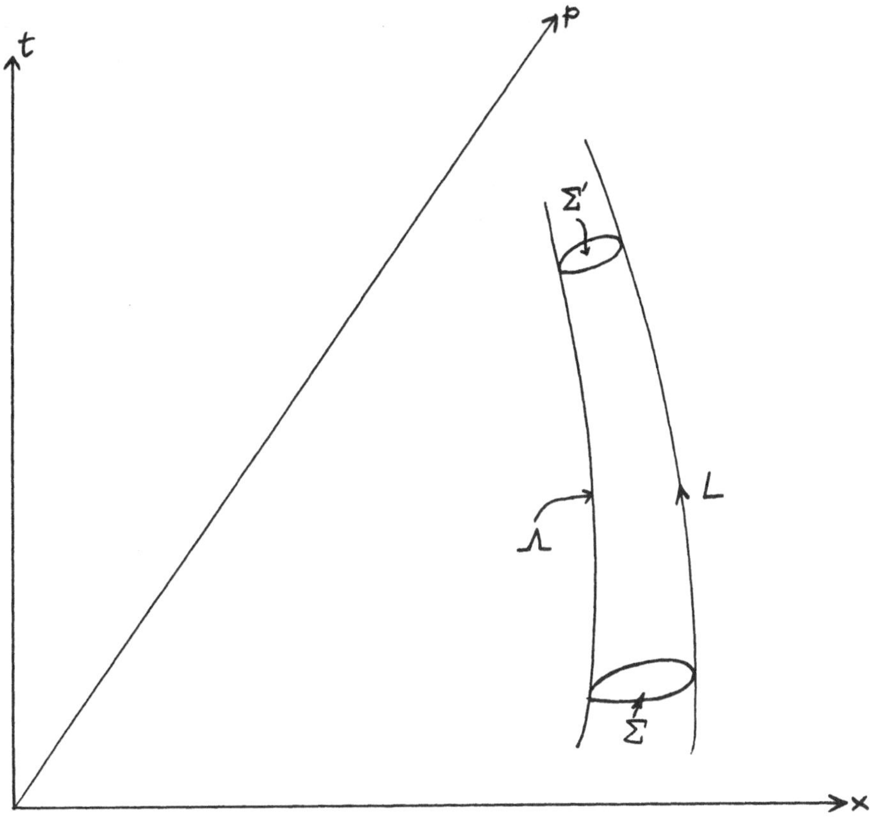

Figure 2. Liouville's theorem. Since $d\omega_m=0$ and $\int\omega_m=0$ for any surface tangent to the worldlines, Gauss' theorem implies that the cross-section of a tube of worldlines has constant ω -measure.

If we choose G small and contained in the local instantaneous rest space of an observer at x with 4-velocity u^a (see fig.3) then $\Sigma = G \times K_x$ is a part of the ordinary phase space associated with that observer with space projection G and momentum domain $K_x \subset P_m(x)$. If K_x is small and contains p (2.14) gives,

$$\omega_m(\Sigma) = p^0 V(G)\, \pi_m(K_x) = p^0 V \pi / p^0 , \qquad (2.15)$$

where $V = V(G)$ is the volume of G and $\pi := \int_{K_x} d^3p$ is the "ordinary" momentum volume of K_x. Hence,

$$\pi_m(G \times K_x) = V\pi ,$$

the ordinary phase space volume the observer would assign to $G \times K_x$. This result shows that the normalisation factor in π_m has been suitably chosen.

Although we shall not make explicit reference to this point again it is highly instructive to consider the relationship between the classical and relativistic descriptions of a gas in phase space. It will usually be found that the relativistic phase space, thanks to its extra dimension, gives a much clearer description of the history of the gas.

2.4 The Distribution Function

We now consider a gas consisting of particles of several types. Particles of type A have rest mass m_A and charge e_A. Microscopic particles may possibly possess other characteristics, eg. baryon number b_A, spin s_A, e- and μ-lepton numbers l_A^e, l_A^μ etc., while macroscopic particles such as stars and galaxies may be further characterised by spectral or galactic types etc. With each type of particle we associate a phase space $M_A := M_{m_A}$ and a Liouville operator L_A. It is obvious that all of the considerations of the previous

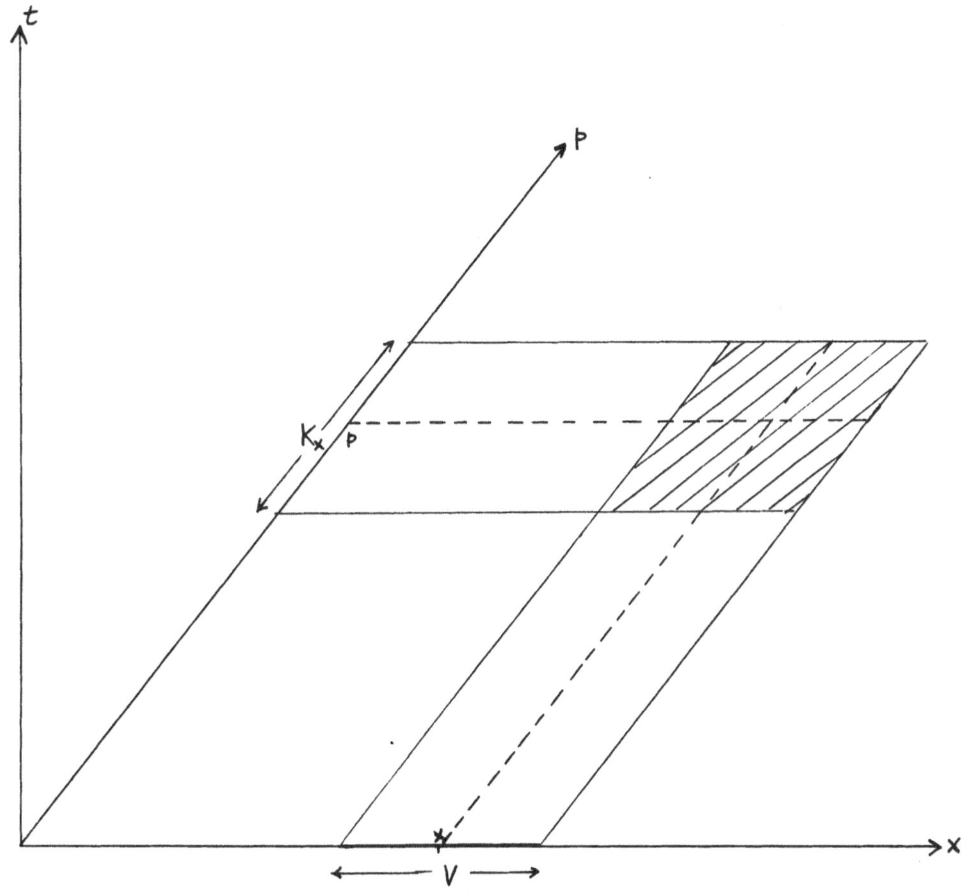

Figure 3. In this 3-dimensional representation of phase space the box G is represented by a spacelike line segment of length V . The momenta of the particles in G lie in the range K_x. The volume occupied by the particles on a spacelike hypersurface t = 0 in phase space is VK_x just as in the classical theory.

sections hold separately for each type of particle. The external or mean fields g_{ab}, F_{ab} are of course independent of A.

A definite history or microstate of the gas can be represented as a collection of segments of phase orbits in M_A the states occupied by A-particles between collisions. A collision will be written symbolically as,

$$(x;\ p_A, p_B \longrightarrow p_C, p_D).$$

This indicates that at x X,particles of types A,B with 4-momenta p_A, p_B collide to produce particles of types C,D with 4-momenta p_C, p_D. The collision gives rise in M_A to an endpoint (x, p_A) of an occupied orbit segment, an annihilation, and in M_C it gives rise to an initial point of such a segment, a creation. More complex collisions including absorption and emission processes are dealt with in an analogous way.

Let Σ be a hypersurface in M_A. (Here a hypersurface means a compact oriented 6-dimensional submanifold with boundary in M_A). Let $N_A[\Sigma]$ be the algebraic number of occupied states intersecting Σ. (An occupied state is counted positively if L_A is directed from the negative side to the positive side of Σ at the point of intersection, negatively if is directed in the opposite direction, and is not counted if L_A is parallel to Σ) The functionals $N_A : \Sigma \rightarrow N_A[\Sigma]$ for all A completely specify the microstate of the gas. It is easy to see that if D is a region in M_A with boundary ∂D then $N_A[\partial D]$ is the number of collisions in D ie. the number of endpoints of occupied segments, if creations are counted positively and annihilations are counted negatively.

We now consider a Gibbs ensemble of microstates which represents a macrostate of the many particle system. We shall henceforth restrict ourselves to the average properties of such an ensemble, as is usual in kinetic theory. These average properties of the ensemble may well provide an approximate macroscopic description of the gas even if the gas obeys quantum laws. This is certainly true in the non-relativistic case where classical (non-quantum) and quantum statistical mechanics often give the same results, (see eg. Kadanoff & Baym (1962)). The ensemble average of $N_A[\Sigma]$ will be denoted by $\overline{N_A}[\Sigma]$

We now make the usual smoothness assumptions:

1) On any fixed hypersurface $\Sigma \subset M_A$ there exists a continuous nonnegative density function $f_{A\Sigma}$ such that for all hypersurfaces $\Sigma' \subset \Sigma$,

$$\overline{N_A}[\Sigma'] = \int_{\Sigma'} f_{A\Sigma}\,\omega \;.$$

The equivalent assumption in the classical case is the existence of a continuous 1-particle distribution function.

2) For any sequence of regions $\{D_i\}$ in M_A which shrinks to a point, there exists a positive number E such that,

$$\left|\overline{N_A}[\partial D_i]\right| \leq E \left|\int_{D_i} \Omega_A\right| .$$

This property implies that the average number of collisions in a region D in M_A is of order at most the Ω-measure of D .

From these two assumptions we can deduce the existence of an <u>invariant 1-particle distribution function $f_A(x,p)$ on M_A</u> such that,

$$\overline{N_A}[\Sigma] = \int_\Sigma f_A\,\omega_A \qquad \text{for any } \Sigma \subset M_A, \tag{2.16}$$

ie. $f_A(x,p)$ is the same for all observers at (x,p) . (A detailed derivation is given in Ehlers (1971). The average number of collisions in the region can now be evaluated by means of the following easily established,

Lemma

If f is any function defined on M,

$$df \wedge L.\Omega = L(f)\Omega \quad .$$

(The details of the proof are given in Ehlers (1971))

The average number of collisions in the region $D \subset M_A$ is,

$$\bar{N}_A[\partial D] = \int_{\partial D} f_A \omega_A ,$$

$$= \int_D d(f_A \omega_A), \text{ by Stokes' theorem,}$$

$$= \int_D df_A \wedge \omega_A, \text{ since } d\omega_A = 0,$$

$$= \int_D df_A \wedge L_A.\Omega_A, \text{ from the definition of } \omega_A,$$

$$= \int_D L_A(f_A)\Omega_A, \text{ by the above lemma.}$$

If we denote the phase space density of collisions by $C_A[f_A]$, then clearly,

$$L_A(f_A) = p^a \frac{\partial f_A}{\partial x^a} + (e_A F^\nu{}_b p^b - \Gamma^\nu{}_{bc} p^b p^c) \frac{\partial f_A}{\partial p^\nu} = C_A[f_A] \quad . \tag{2.17}$$

Equation (2.17) would describe the evolution of the 1-particle distribution function f_A provided we know the functional $(f_A, f_B, \ldots) \rightarrow C_A[f_A]$ from an analysis of the collision mechanism. This problem will occupy the next section.

2.5 The Boltzman Equation.

Since we wish to obtain equations which describe the evolution of f_A with time it is necessary to give a prescription for estimating the average number of collisions in a small spacetime region D with momenta lying in prescribed ranges in the appropriate mass shells. In a rigorous many particle theory one would expect this number to be a functional of the multiparticle distribution functions, and equations describing the evolution of these would be required, as in non-relativistic statistical mechanics. But as yet no relativistic analogue of theBBGKY hierarchy has been rigorously presented, and instead we follow the simple approach of Boltzmann in which this number is approximated in terms of the 1-particle distribution functions. The underlying physical assumptions are as follows:

1) the particles interact only through,

   a) very long range forces which can be approximately accounted for by mean field tensors such as g_{ab}, F_{ab},

   b) very short range forces such as hard core interactions whose effect can be approximated by instantaneous collisions, annihilation and creation processes.

2) the system is not too far from equilibrium,

3) the gas is not too cold and dense so that particles which are about to collide have uncorrelated momenta.

The relativistic analogue of Boltzmann's "Stosszahlansatz" can be presented for a special relativistic gas enclosed in a cubical box G. We regard the spacetime region D as the product of the spacelike hypersurface (box) G' with a time interval of duration T. It is assumed that T and G are small so that D can be regarded as flat as far as the collisional behaviour of the gas in D is concerned, and that F_{ab} is sufficiently weak not to affect the collisions. Let u^a be the 4-velocity of G. If the average number of collisions within D of the type,

$$(x; p_A, p_B, \ldots \longrightarrow p_C, p_D, \ldots)$$

with $p_A \in K_A \subset M_A$ etc. is $\mathcal{N}$ and if we assume that the interaction Hamiltonian is spin-independent, then it can be shown (see eg. Ehlers (1971)) that, in first order perturbation theory, and with some simplifying assumptions,

$$\mathcal{N} = (TV)\,\pi_A(K_A)\,\pi_B(K_B)\ldots\pi_C(K_C)\,\pi_D(K_D)\ldots f_A f_B \ldots \times \\ \times\ \hat{f}_C \hat{f}_D \ldots W(x; p_A, p_B, \ldots \rightarrow p_C, p_D, \ldots), \tag{2.18}$$

where $f_A := h^3/r_A f_A(x,p_A)$, etc., $\hat{f}_C := 1 \pm h^3/r_C f_C(x,p_C)$ and the upper sign always refers to bosons, while the lower sign always refers to fermions. The factor h is Planck's constant and r_A is the spin degeneracy of particles of species A. We can always choose units to set h equal to 1, and we shall nearly always omit these factors in subsequent calculations. The expression (2.18) has been obtained from first order Dirac perturbation theory, and so $\mathcal{N}$ contains a factor $\delta(p_A+p_B+\ldots-p_C-p_D-\ldots)$ which has been incorporated into the W term. We shall assume that W is a Lorentz invariant "function" of its momentum variables, independent of the position and size V of the quantization volume G. It follows from Dirac perturbation theory that to this approximation, W is invariant under combined space and time reflections (PT),

$$W(p_A,p_B,\ldots \rightarrow p_C,\ldots) = W(p_C,\ldots \rightarrow p_A,p_B,\ldots). \tag{2.19}$$

The argument leading to the expression (2.18) contains several somewhat obscure approximations. An alternative, and in some ways more satisfactory treatment is given in the appendix to chapter 2.

We can now obtain an expression for the time development of f_A by adding together expressions of the form (2.18) for all of the collisions in which A-particles participate. We consider a fairly general case in which collisions occur between particles of different species; binary collisions (both elastic and inelastic,) and emission and absorbtion processes are allowed. Then the equation governing the evolution of f_A, the generalised Boltzmann Equation, is,

$$\begin{aligned} L_A(f_A) = \tfrac{1}{2} \sum_{B,C,D} \iiint (\hat{f}_A \hat{f}_B f_C f_D - f_A f_B \hat{f}_C \hat{f}_D) W_{AB\rightarrow CD} \pi_B \pi_C \pi_D \; + \\ + \sum_{B,C} \iint (\hat{f}_A \hat{f}_B f_C - f_A f_B \hat{f}_C) W_{AB\rightarrow C} \pi_B \pi_C \; + \\ + \tfrac{1}{2} \sum_{BC} \iint (\hat{f}_A f_B f_C - f_A \hat{f}_B \hat{f}_C) W_{BC\rightarrow A} \pi_B \pi_C \; , \end{aligned} \tag{2.20}$$

where the symmetry condition (2.19) has been used. In these integrals we write $\pi_B \pi_C$ for $\pi_B \wedge \pi_C$ etc. This equation has been previously derived by many authors including Lichnerowicz & Marrot (1940), Tauber & Weinberg (1961), Chernikov (1963), Israel (1963), Bichteler (1965), Marle (1969), Ehlers (1971), Sachs (1971). The derivation is standard and follows the classical theory. The details are given in Ehlers (1971). The factors ½ appear to ensure that each collision is only counted once, and the integrals are taken over the whole of the appropriate phase spaces. In chapter 5 where we deal with collision integrals, W which is a measure of the rate at which collisions occur, will be related to the more useful differential cross section.

2.6 Moments of the Distribution Function, Entropy, the H-theorem and Equilibrium.

In subsequent chapters we shall need to use the moments of $f_A(x,p)$. These are defined as follows:

$$A_A(x) := \int f_A \pi_A, \tag{2.21}$$

$$N_A{}^a(x) := \int p^a f_A \pi_A, \tag{2.22}$$

$$T^{ab}{}_A(x) := \int p^a p^b f_A \pi_A, \tag{2.23}$$

$$S^{abc}{}_A(x) := \int p^a p^b p^c f_A \pi_A, \tag{2.24}$$

$$Q^{abcd}{}_A(x) := \int p^a p^b p^c p^d f_A \pi_A, \tag{2.25}$$

$$P^{abcde}{}_A(x) := \int p^a p^b p^c p^d p^e f_A \pi_A, \tag{2.26}$$

where the integrals are over $P_A(x)$. $N^a{}_A(x)\sigma_a$ is the flux of particles across a 3-surface element σ_a, and so $N_A{}^a(x)$ is the flux density. In a similar way $T_A{}^{ab}(x)$ is the momentum flux density tensor, or energy-momentum tensor. A similar interpretation can be given for the higher order moments. We now prove an important lemma that will be required in subsequent chapters.

Lemma

$$N_A{}^a{}_{;a}(x) = \int L_A(f_A)\,\pi_A, \qquad (2.27)$$

$$T_A{}^{ab}{}_{;b}(x) = \int p^a L_A(f_A)\,\pi_A, \text{ if } F^{ab} = 0, \qquad (2.28)$$

with corresponding results for higher order moments.

Proof

Let V an arbitrary region in spacetime X and let D be the corresponding region in phase space M_A. Then,

$$\int_V \eta\, N_A{}^a{}_{;a}(x) = \int_{\partial V} N_A{}^a \sigma_a, \quad \text{by Gauss' theorem, (2.9),}$$

$$= \int_{\partial V} \int_{P_A(x)} f_A(x,p)\, p^a \sigma_a \wedge \pi_A, \quad \text{from (2.22),}$$

$$= \int_D f_A \omega_A, \quad \text{by (2.14),}$$

$$= \int_D L_A(f_A)\,\pi_A, \quad \text{by the argument leading to (2.17),}$$

$$= \int_V \int_{P_A(x)} L_A(f_A)\,\pi_A, \quad \text{by (2.12).}$$

Since this holds for arbitrary V,

$$N_A{}^a{}_{;a}(x) = \int_{P_A(x)} L_A(f_A)\,\pi_A,$$

which proves (2.27), and (2.28) is proved in a similar way,

The entropy flux vector for the A-component of the gas is defined by analogy with quantum statistical mechanics to be,

$$S_A{}^a(x) := -k \int_{P_A(x)} (f_A \log f_A \mp f_A \log f_A)\, p^a\, \pi_A. \qquad (2.29)$$

The total entropy flux vector of the gas is defined to be,

$$S^a(x) := \sum_A S_A{}^a(x). \qquad (2.30)$$

Just as in the classical theory there is a H-theorem, (see eg. Tauber & Weinberg (1961), Ehlers (1961), Chernikov (1963))

The H-Theorem

The total entropy flux vector $S^a(x)$ satisfies the inequality

$$S^a{}_{;a}(x) \geqslant 0. \qquad (2.31)$$

Proof

(2.29), (2.30) imply, using the argument of the lemma that,

$$S^a{}_{;a} = \sum_A k \int \log(\hat{f}_A/f_A) L_A(f_A)\, \pi_A,$$

$$= \tfrac{1}{2}k \sum_{A,B,C,D} \iiiint \log(\hat{f}_A/f_A)(\hat{f}_A \hat{f}_B f_C f_D - f_A f_B \hat{f}_C \hat{f}_D) W_{AB\to CD}\, \pi_A \pi_B \pi_C \pi_D +$$

+ 2 similar terms,

where the Boltzmann equation (2.20) has been used. The first term on the right hand side is unaltered if we interchange A,B throughout, and changes sign if we exchange A,B for C,D. Similar rules apply to the other two terms. Therefore,

$$S^a{}_{;a} = 1/8\, k \sum_{A,B,C,D} \iiiint \left\{ \log(\hat{f}_A/f_A) + \log(\hat{f}_B/f_B) - \log(\hat{f}_C/f_C) - \log(\hat{f}_D/f_D) \right\} \times$$

$$\times (\hat{f}_A \hat{f}_B f_C f_D - f_A f_B \hat{f}_C \hat{f}_D) W_{AB\to CD}\, \pi_A \pi_B \pi_C \pi_D +$$

+ 2 similar terms,

or,

$$S^a{}_{;a} = 1/8\, k \sum_{A,B,C,D} \iiiint \log(\hat{f}_A \hat{f}_B f_C f_D / f_A f_B \hat{f}_C \hat{f}_D) \times$$

$$\times (\hat{f}_A \hat{f}_B f_C f_D - f_A f_B \hat{f}_C \hat{f}_D) W_{AB \to CD} \pi_A \pi_B \pi_C \pi_D +$$

$$+ 2 \text{ similar terms.}$$

Each of the three terms on the right hand side is of the form,

$$\sum \iiiint (\log F - \log G)(F-G) W_{AB \to CD} \pi_A \pi_B \pi_C \pi_D ,$$

and hence is necessarily non-negative. This proves the theorem.

A collisional invariant is a function $\varphi_A(x,p)$ defined on each phase space such that,

$$\begin{cases} \varphi_A + \varphi_B - \varphi_C - \varphi_D = 0 & \text{for binary collisions } (x; p_A, p_B \to p_C, p_D), \\ \varphi_A + \varphi_B - \varphi_C = 0 & \text{for annihilations } (x; p_A, p_B \to p_C), \\ \varphi_A - \varphi_B - \varphi_C = 0 & \text{for creation processes } (x; p_A \to p_B, p_C). \end{cases}$$

In physical terms a collisional invariant is a quantity which is conserved in all collisions and creation/annihilation processes. Examples of collisional invariants include the 4-momentum p^a, baryon number b_A, and, if only binary collisions are involved, 1.

In fact it can be shown, (see eg. (Chernikov (1963) Bichteler (1965), Boyer (1965), Marle (1969), Ehlers (1971)), that these are essentially all of the collisional invariants in the following sense: the most general collisional invariant φ_A can be expressed in the form

$$\varphi_A = \beta_a p^a + \alpha_A, \tag{2.32}$$

where $p^a \in P_A(x)$, $\beta_a = \beta_a(x)$ is independent of A, and α_A is a linear combination of the scalars conserved in collisions, ie. charge, baryon number, lepton number, spin etc. where appropriate. The proof is intricate and will be found in the cited references.

We shall define an equilibrium state of the gas as one in which $S^{a}{}_{;a} = 0$. It follows from the proof of the H-theorem that an equilibrium state is only possible if one of the following two conditions is satisfied:

i) $W = 0$, that is there are no collisions,

ii) $\log(\hat{f}_A/f_A)$ is a collisional invariant.

In both cases,

$$L_A(f_A) = 0 \quad \text{for all } p^a \in P_A(x), \text{ all } A. \tag{2.33}$$

Condition (2.33) states that f_A is a constant of the motion either because there are no collisions, or because the net effect of collisions is to leave f_A unaltered. This latter case is termed detailed balancing. From condition ii) and (2.32) we can obtain the most general form for a collision dominated equilibrium distribution function, the relativistic Bose-Einstein & Fermi-Dirac distributions,

$$\frac{h^3}{r_A} f_A(x,p) = \left[\exp(-\alpha_A(x) - \beta_a(x)p^a) \mp 1\right]^{-1}, \tag{2.34}$$

where $\beta_a(x)$ is independent of A, and the $\alpha_A(x)$ depend on particle type as described above. Here the factor h^3/r_A required to give f_A the right dimensions has been explicitly included. It can easily be verified that the moments of f_A given by (2.21)-(2.26) will only be finite if $\beta_a(x)$ is a timelike future directed vector field,

$$\beta_a(x) = \beta(x)u_a(x), \quad u_a u^a = -1, \quad \beta > 0, \quad u^0 > 0. \tag{2.35}$$

Physically this means that f_A tends to zero as the energy tends to infinity, a reasonable requirement. If we now substitute (2.35) and (2.34) into the expression for S^a and compare the resulting expression with its thermostatic counterpart, (given in the next chapter,) we may deduce that,

$$\beta = 1/kT, \qquad \alpha_A = \kappa_A/kT, \tag{2.36}$$

where k is Boltzmann's constant, T is the temperature of the mixture, and κ_A is the chemical component of the A-component in the mixture. It turns out that the 4-velocity $u^a(x)$ introduced in the decomposition of β_a is parallel to the total flux density, so that we may interpret it as the mean fluid velocity. Thus all of the parameters in the formula (2.34) for the collision dominated equilibrium distribution function have thermodynamic significance. (A collision dominated equilibrium distribution is one in which equilibrium is preserved by detailed balancing.) In the classical case (2.34) reduces to the relativistic Maxwell-Boltzmann distribution,

$$\frac{h^3}{r_A} f_A(x,p) = \exp(\alpha_A(x) + \beta_a(x)p^a), \tag{2.37}$$

where the dimensional factor h^3/r_A has been explicitly included. Suppose we consider a mixture of gases in equilibrium. Then, considering a collision of the type $(x; p_A, p_B \rightarrow p_C, p_D)$ it follows from equation (2.32) that

$$\alpha_A + \alpha_B - \alpha_C - \alpha_D = 0,$$

with similar results for annihilation and creation processes, From equation (2.36) we may then deduce,

$$\mathcal{A} := \kappa_C + \kappa_D - \kappa_A - \kappa_B = 0.$$

Either of these two equations may, following Marle (1969), be regarded as the relativistic formulation of the law of mass action. For a classical derivation see eg. de Groot & Mazur (1962), Meixner (1958). If we normalise the stoichiometric coefficients for such a reaction to ± 1 then $\mathcal{A}$ defined above is called the affinity of the reaction. In equilibrium the affinities for each reaction vanish.

In view of their comparative simplicity it is important to consider what conditions are imposed on the underlying spacetime by the assumption of equilibrium in the phase space. The collision dominated distributions are examples of a more general class, the locally dynamically symmetric distributions. A distribution of particles is said to be locally dynamically symmetric (LDS) if there exists a timelike unit vector field $u_a(x)$ such that at each point $x \in X$ $f_A(x,p)$ is invariant with respect to all of the homogeneous Lorentz transformations in the tangent (phase) space which leave u^a unchanged. Analytically this means that has the form,

$$f(x,p) = h(x, u_a(x)p^a),$$

so that the physical meaning is that the distribution function with respect to an observer comoving with the gas, depends only on position and energy of the particles. We now introduce some more definitions, (see eg. Ehlers (1961),):

$h_{ab} := u_a u_b + g_{ab}$ is the projection tensor into the rest space of

$\theta := u^a{}_{;a}$ is the scalar expansion rate,

$\dot{u}^a := u^a{}_{;b} u^b$ is the acceleration vector,

$\omega_{ab} := u_{[a;b]} + \dot{u}_{[a} u_{b]}$ is the vorticity tensor,

$_{ab} := u_{(a;b)} + \dot{u}_{(a} u_{b)} - \frac{1}{3} h_{ab} \theta$ is the shear tensor,

$\omega := (\frac{1}{2} \omega_{ab} \omega^{ab})^{\frac{1}{2}}$ is the vorticity scalar.

The main results for LDS distributions are given in the following theorem: (it is implicitly assumed that the gas has only one component, but this assumption is only made for simplicity.)

<u>Theorem</u>

If spacetime contains a gas consisting of a locally dynamically symmetric distribution of particles, then,

a) the gas is shearfree,

b) the spatial projection of its acceleration is proportional to the gradient of a scalar,

c) it cannot be simultaneously rotating and expanding, ie. $\omega\theta = 0$.

<u>Proof</u>

Since the proof of part c) is intricate we shall only give the details for parts a), b). By assumption the gas defines a unit timelike vector field $u^a(x)$ and the particles have an isotropic distribution in the rest frame of u^a. Therefore the moments of f can pick out no preferred direction in X other than those determined by u^a and g^{ab}. Consequently the only form for the second moment is,

$$T^{ab}(x) = X(x)u^a u^b + Y(x)g^{ab}$$

which shows that the gas must behave like a perfect fluid. By a similar argument the third moment must be of the form,

$$S^{abc}(x) = U(x)u^a u^b u^c + 3V(x)u^{(a}g^{bc)}. \qquad (2.38)$$

Just as in (2.27), (2.28) we can express $S^{abc}{}_{;c}$ as the second moment of $L(f)$ and using the Boltzmann equation we can express $S^{abc}{}_{;c}$ in terms of an integral involving $C[f]$. It follows that $S^{abc}{}_{;c}$ must have the same form as T^{ab} but with different scalar functions in place of X,Y. Differentiating (2.38) gives,

$$S^{abc}{}_{;c} = (\dot{U}+U\theta+\tfrac{2}{3}V\theta)u^a u^b + (\dot{V}+5/3V\theta)g^{ab} + 2V\sigma^{ab} + $$
$$+ 2u^{(a}\left[U\dot{u}^{b)} + V^{,b)} - V\dot{u}^{b)}\right] ,$$

from which may be concluded,

$$\sigma_{ab} = 0, \qquad \dot{u}^a = -h^{ab}V_{,b}/(U-V),$$

which proves parts a), b). Part c) has been proved partly by Ehlers, Geren & Sachs (1968), and partly by Robert Treciokas (1971). The theorem can easily be extended to the case of a mixture of gases if the distributions of the individual components are all isotropic in some reference frame. This would be the case if the gas mixture was in collision dominated equilibrium.

For a gas in collision dominated equilibrium there is the further requirement that the left hand side of the Boltzmann equation should vanish,

$$L_A(f_A) = 0, \text{ all } p^a \in P_A(x), \text{ all } A. \qquad (2.39)$$

Applying this condition to (2.34) or (2.37) gives,

$$(\alpha_{A,a} + e_A F^b{}_a \beta_b)p^a + \beta_{(b;c)}p^b p^c = 0. \qquad (2.40)$$

(2.40) has to hold for all $p^a \in P_A(x)$ and so we may deduce, if $m \neq 0$,

$$(\alpha_{A,a} + e_A F^b{}_a \beta_b) = 0, \qquad \beta_{(b;c)} = 0. \qquad (2.41)$$

The antisymmetry of $F^b{}_a$ implies $\dot{\alpha}_A := \alpha_{A,a}u^a = 0$ so that in the rest frame of u^a, α_A appears to be constant. If the electric field as measured in the same frame is zero, then $\alpha_{A,a} = 0$ and α_A is constant. If $m_A = 0$ then the second of the equations (2.41) cannot be deduced from equation (2.40). If $m_A \neq 0$, β^a is a timelike Killing vector, and so X cannot admit expansion. Thus we have proved,

<u>Theorem</u>

Suppose spacetime X contains a mixture of gases in collision dominated equilibrium. Then the $\alpha_A(x)$ the ratios of the chemical potentials to the temperature, have no time dependence measured in the rest frame of the gas mixture.

If the electric field vanishes in this frame then the $\alpha_A(x)$ have no spatial dependence. The gas motion is always shearfree, and if one of the components has non-zero rest mass, only rigid(non-expanding, non-shearing, possibly rotating) motion is possible, and spacetime is stationary.

This theorem supplies part of the motivation for the study of non-equilibrium relativistic kinetic theory. If. as in cosmology, we are concerned with non-stationary solutions of Einstein's field equations then the simple equilibrium distributions (2.34), (2.37) are only applicable to an isotropically expanding gas whose particles all have zero rest mass. Such a model could be used to represent the early stages of the universe in a "hot big bang" model. But if we wish to include shear as well as expansion, or to consider an expanding gas, the particles of which have non-zero rest mass, or to consider an expanding rotating gas, (if such solutions are possible,) then it is necessary either to consider non-equilibrium distributions, or to consider collisionfree equilibrium distributions. Most of this work will be taken up with non-equilibrium distributions, and so we conclude this chapter with a brief discussion of collisionfree distributions, which are of course necessarily equilibrium ones.

A collisionfree distribution function must satisfy,

$$L_A(f_A) = 0, \text{ all } p^a \in P_A(x), \text{ all } A. \qquad (2.33)$$

Since L_A is the directional derivative along the particle trajectories, (2.33) implies that f_A is a constant of the motion. If f_A is given on an initial spacelike hypersurface, and sufficiently many constants of the motion are known, then f_A is determined at all points in M_A. Since 1965 several workers have become interested in this problem including Zel'dovich & Podurets (1965), Fackerell (1966,1968), although the original theory of the collisionless Boltzmann equation dates back to Synge (1934) and Walker (1936). Most of the work has been on stationary spherically symmetric systems such as star clusters. If spacetime admits a symmetry map this map induces a transformation in M_A and if we require f_A to be invariant under this induced map, f_A can only depend on those constants of the motion which are invariant under the symmetry map. In the high symmetry configurations normally studied, f_A only depends on one or two parameters, and if the functional form of f_A is chosen, (usually a truncated Maxwellian distribution,) the coupled Boltzmann-Einstein equations can be solved, at least in principle.

Less work has been published for non-stationary systems. The basic difficulty is in constructing constants of the motion. In the case of neutral particles f_A must be constant on all timelike or null geodesics. Now if $k^a(x)$ is a Killing vector field $k^a(x)p_a$ is constant along geodesics since,

$$\frac{d}{dv}(k_a p^a) = k_{a;b}p^a p^b = k_{(a;b)}p^a p^b = 0. \quad .$$

There are of course other methods of generating constants of the motion, see eg. Carter (1969), Walker & Penrose (1970) but we shall not consider them here.

Suppose spacetime has symmetries generated by the Killing vectors $k^a_\alpha(x)$ $\alpha = 1,2,\ldots,r.$ If g is any positive function of r variables, $f(x,p) = g(k^a_\alpha(x)p_a)$ is a solution of the collision-free Boltzmann equation. If f is to be invariant under the infinitesimal transformations generated by the k^a_α it is easy to show (Ehlers (1971)) that

$$\frac{\partial g}{\partial y_\alpha} C^\gamma_{\alpha\beta}\, y_\gamma = 0,$$

where $y_\alpha := k^a_\alpha p_a$ and $C^\gamma_{\alpha\beta}$ are the structure constants of the symmetry group,

$$[k_\alpha, k_\beta] = C^\gamma_{\alpha\beta}\, k_\gamma \quad .$$

Since this equation is linear it can be solved for a specified group, and solutions of the collisionless Boltzmann equation invariant under groups of spacetime isometries can be obtained.

Appendix to Chapter 2: Alternative Derivation of the Boltzmann Equation.

The derivation of the Boltzmann equation given in chapter 2, based on the original Boltzmann "Stosszahlansatz", contains several somewhat obscure approximations, and it is therefore worthwhile to consider whether a more satisfactory derivation is possible. The usual treatments involve a detailed study of the dynamics of an N-particle system in which the dynamical variables are, in principle, exactly measurable (apart from quantum limitations). However the exact state of the gas is never known, (although in principle it could be,) and so the resulting equations are averaged over a Gibbs ensemble of possible states. It is clear from the discussion in the introduction that there will be severe difficulties involved in extending this discussion to the relativistic case.

Recently however Penrose (1970) has suggested an alternative approach in which the dynamical observables are to be regarded as random variables from the start. In this probabilistic approach, the dynamics of the system are subservient to the statistical analysis. The basic assumption is that measurements of certain dynamical observables at successive instants form a Markov chain. In principle this assumption is capable of verification by experiment. In practice the assumption is made plausible by an appeal to the dynamical equations of the system. From this and a few other basic postulates, Penrose develops a complete and logically consistent kinetic theory. It is therefore worthwhile examining whether the theory can be generalised to include the effects of relativity.

In any discussion of interacting particles some assumption of causality must be made. Here we shall assume that spacetime is stably causal. It immediately follows, (Hawking(1968)) that there exists a cosmic time function, ie. a family of non-intersecting smooth spacelike hypersurfaces t = constant. These hypersurfaces obviously induce a family of smooth non-intersecting 6-dimensional hypersurfaces Σ in phase space. Each Σ corresponds to the instantaneous phase space of the gas. We choose an arbitrary Σ_0 and partition it into cells of equal (covariant) volume h^3. We can induce unique partitions on all the other Σ by parallely propagating the boundaries of the cells in Σ_0 along the timelike or null geodesics. From Liouville's theorem it follows that all of the cells on all of the Σ have volume h^3.

Suppose that a countable subset of the hypersurfaces are labelled by $A = 0,1,2,\ldots$ and the cells are labelled by $i = 1,2,3,\ldots,$ on each hypersurface, so that the ith cell on the Ath hypersurface can be denoted by ν_{Ai}. Let $n[\nu_{Ai}]$ be the number of particle worldlines crossing ν_{Ai} (As was explained in chapter 2, the functional $n[\nu_{Ai}]$ is related to the distribution function f in the limit $h \to 0$). Because of the obvious limitations on the accuracy and completeness of observations of the system we consider the $n[\nu_{Ai}]$ as random variables which completely describe the system, in the sense that two systems with the same $\{n[\nu_{Ai}]\}$, (which can be dynamically different, ie. distinguishable on a microscopic scale,) cannot be observationally distinguished.

This postulate already makes an assumption about the physics of the system. We can conventionally represent the particle interactions as the sum of three effects,

1) long range interactions represented by mean gravitational and electromagnetic fields which affect the particles via the Einstein-Maxwell equations,

2) short range interactions represented by instantaneous collisions, annihilation and creation processes,

3) intermediate range interactions, not of the above two types.

The $\{n[\nu_{Ai}]\}$ can only give a physically reasonable description of the system if the fields representing the interactions of the third type are reasonably constant on each cell, and these interactions are very small compared with those of the first and second types, so that we may neglect them.

We now make the fundamental assumption that for $A = 0,1,2,\ldots$ the set $\{n[\nu_{Ai}]\}$ form a Markov chain. Although this postulate is in principle verifiable, the experiment would be difficult, and Penrose argues for its plausibility because of the existence of equations of motion for the particles. (For the details see Penrose (1970).) Bearing in mind the assumptions already made above, Penrose's argument should extend to the relativistic case. Considering only the case of a 1-component classical gas Penrose obtains a Master equation for the Markov chain in terms of the elastic collision cross-sections. By taking the limit $h \to 0$ and the transition from the $n[\nu_{Ai}]$ to $f(x,p)$ he obtains the usual Boltzmann equation with an error term of order $1/N$ x the normal Boltzmann collision term, where N is the total number of particles. Such a calculation does go through in the relativistic case. The details are left as an exercise for the reader. Clearly the final equation is independent of the choice of partition of the $\{\Sigma\}$ and indeed is independent of the choice of the $\{\Sigma_i\}$. Thus the final equation is covariant.

Although such a derivation is both concise and fairly rigorous, it is conceptually quite difficult, and, unlike equations derived from a BBGKY hierarchy, it cannot be generalised to the case of a dense gas.

CHAPTER 3. RELATIVISTIC THERMODYNAMICS OF IRREVERSIBLE PROCESSES.

1. Introduction and Definitions.

Most of this work is concerned with the kinetic theory description of non-equilibrium processes in relativistic gases. This involves some fairly complicated mathematics and so we first give a relatively simple description of the effects by presenting a phenomenological treatment of irreversible processes within the framework of general relativity. It is the straightforward generalisation of the special relativistic theory of Kluitenberg & de Groot (1953), Meixner (1958) except for a few alterations. An alternative treatment has been given by Stueckelberg & Wanders (1953) which is somewhat less general than the theory presented here, but is otherwise equivalent. The main reasons for including such a discussion here are:

a) it gives a simple and straightforward physical picture of non-equilibrium processes,

b) it suggests the form of the transport equations and relativistic Casimir-Onsager reciprocal relations, (which of course have ultimately to be verified by kinetic theory or statistical mechanics,) and these are often sufficient for theoretical calculations where numerical values are not required.

Let the particle flux density of the A-th component, $(A = 1,2,\ldots,N)$, in an N-component fluid be,

$$N_A{}^a = n_A u_A{}^a \, , \text{ where } u_{Aa} u_A{}^a = -1. \tag{3.1}$$

In the rest frame of u_{Aa}, n_A is the number density, ie. the number of particles per unit measured, (not proper) volume.

Let c_A denote scalars which are conserved for each particle of the A-species, eg. baryon number, lepton number, charge, and in certain cases, rest mass, number etc. We denote the total <u>c -flux density</u> by

$$N_c{}^a := \sum_A c_A N_A{}^a = n_c u_c{}^a \text{ say, where } u_{ca} u_c{}^a = -1. \quad (3.2)$$

$u_c{}^a$ is the mean velocity of the fluid, (with respect to c_A) and n_c is the total c -density measured in the rest frame of $u_c{}^a$.

Because c_A is a conserved quantity we have the conservation law,

$$\sum_A (c_A N_A{}^a)_{;a} = 0, \quad (3.3)$$

or,

$$N_c{}^a{}_{;a} = 0. \quad (3.4)$$

A case of particular interest is when $c_A = m_A$, the rest mass. $u_c{}^a$ is then called the <u>barycentric velocity</u>, and n_c is usually written as ρ the <u>rest mass density</u>.

We now consider an observer with 4-velocity u^a (which can, but need not be, the barycentric velocity.) Equations which hold only in the rest frame of the observer will be prefixed by a star, eg.

$$* \quad u^a = (1,0,0,0) \quad (3.5)$$

Using the operator $h^{ab} := u^a u^b + g^{ab}$ which projects into the rest space of u^a we may write,

$$N_A{}^a = n_A u_A{}^a + J_A{}^a, \quad (3.6)$$

where ,

$$J_A{}^a = h^{ab} N_{Ab}. \quad (3.7)$$

n_A is the observed number density of A-particles, and $J_A{}^a$ is their observed <u>diffusion flux</u>. From (3.7) we have,

$$u_a J_A{}^a = 0, \qquad (3.8)$$

and if $u^a = u_c{}^a$,

$$\sum_A c_A J_A{}^a = 0. \qquad (3.9)$$

Let Ω_A be the A- production density, a scalar defined by,

$$\Omega_A := N_A{}^a{}_{;a}. \qquad (3.10)$$

Since c_A is conserved then,

$$\sum_A c_A \Omega_A = 0, \qquad (3.11)$$

which is equivalent to,

$$N_c{}^a{}_{;a} = 0. \qquad (3.12)$$

We can write the total energy-momentum tensor in the form,

$$T^{ab} := (\mu + p)u^a u^b + pg^{ab} + 2\, q^{(a}u^{b)} + \pi^{ab}, \qquad (3.13)$$

where μ is the energy density, p the isotropic kinetic pressure, q^a the heat flow relative to u^a, $(q_a u^a = 0,)$ and π^{ab} is the tracefree stress tensor, ($\pi^{ab}u_b = \pi^a{}_a = 0$), all quantities being measured in the rest frame of u^a. If the gravitational field equations of general relativity are to be satisfied,

$$T^{ab}{}_{;b} = 0. \qquad (3.14)$$

If certain components do not interact with the rest of the fluid, then (3.14) is satisfied for each of their energy momentum tensors separately.

3.2 The Gibbs Equation.

If we omit such substances as superfluids from the discussion, and consider only small deviations from equilibrium, then in certain frames in local Euclidean coordinates it is reasonable to assume that the Gibbs equation (see eg. de Groot & Mazur (1962),)holds locally,

$$* \qquad T\,d(Vs) = d(\mu V) + p^* dV - \sum_A \kappa_A d(n_A V), \qquad (3.15)$$

(note the star preceding the equation,) where s is the entropy per unit volume (entropy density), p^* is the static (thermodynamic) pressure, and κ_A is the chemical potential of the A-substance in the mixture, all quantities being measured in the observer's frame. In equilibrium this relation is supposed to exist between the quantities, while for small deviations from equilibrium the temperature T, p^* and κ_A can be defined as the appropriate derivatives, as in the classical theory,

$$T := (\partial\mu/\partial s)_{V,n_A}\ ,\ p^* := -(\partial\mu/\partial V)_{s,n_A} + Ts - \mu + \sum_A n_A\kappa_A,\ \kappa_A := (\partial\mu/\partial n_A)_{s,V}. \quad (3.16)$$

Equation (3.16) cannot be put into a covariant form until the transformation properties of the quantities entering into it have been specified. Classical thermodynamics of course gives little indication of the possible generalization because the kinematical effects are assumed to be separable from the thermal ones. If we write "$\cdot$" $= {}_{,a}u^a$ then (3.16) implies,

$$* \qquad V^{-1}T(Vs)^{\cdot} = \dot{\mu} + (\mu + p^*)\theta - \Big(\sum_A \kappa_A (n_A V)^{\cdot}\Big)/V\ . \qquad (3.17)$$

Equation (3.14) implies $u_a T^{ab}{}_{;b} = 0,$ which leads to

$$\dot{\mu} = -(\mu + p)\Theta - \pi_{ab}\sigma^{ab} - q^a{}_{;a} - q^a \dot{u}_a \, , \qquad (3.18)$$

where the shear tensor σ^{ab} and acceleration $\dot{u}^a$ were defined in chapter 2.

From equation (3.17) it is easy to see that if $\Theta \neq 0$, but $(Vs)^\cdot = (Vn)^\cdot = 0$,

$$* \qquad p^* = -\dot{\mu}\Theta^{-1} - \mu \, , \qquad (3.19)$$

so that

$$* \qquad p^* = p + \Theta^{-1}(\pi_{ab}\sigma^{ab} + q^a{}_{;a} + q^a \dot{u}_a) \quad . \qquad (3.20)$$

It will be shown later than these conditions imply $\pi_{ab} = q_a = 0,$ and so

$$* \qquad p^* = p. \qquad (3.21)$$

However (3.21) does not hold in general

From equations (3.6), (3.10)

$$\dot{n}_A = -n_A\Theta - J_A{}^a{}_{;a} + \Omega_A \, ,$$

so that

$$(n_A V)^\cdot / V = - J_A{}^a{}_{;a} + \Omega_A \, . \qquad (3.22)$$

Thus equation (3.17) can be written,

$$* \; (Vs)^\cdot/V = -(\pi_{ab}\sigma^{ab} + q^a{}_{;a} + q^a\dot{u}_a)/T + \pi\Theta/T + \sum_A \kappa_A (J_A{}^a{}_{;a} - \Omega_A)/T, \qquad (3.23)$$

where $\pi := p^* - p$ is the difference between the static and dynamic pressures measured by the observer.

We now define

$$* \qquad X_{Aa} := -(\kappa_A/T)_{;a} \, , \; A = 1,2,\ldots,N, \qquad (3.24)$$

$$* \qquad X_{0a} := (1/T)_{,a} - \dot{u}_a/T \ , \tag{3.25}$$

$$J_{0a} := q_a \ , \tag{3.26}$$

where X_{Aa} is, apart from a trivial constant, the gradient of α_A introduced in chapter 2. The covariant derivatives are purely formal since (3.24), (3.25) hold only in the observer's frame in locally Euclidean coordinates until we specify the transformation properties of the thermodynamic variables. Suppose that there are n independent chemical reactions occurring in the mixture. Let ν_{Ar} be the stoichiometric coefficient of the A-th component in the rth reaction, normalised as in chapter 2; eg. if the rth reaction is

$$A + B + 2C \rightarrow D + E \ ,$$

then $\nu_{Ar} = \nu_{Br} = -1$, $\nu_{Cr} = -2$, $\nu_{Dr} = \nu_{Er} = +1$. For each A we may write

$$\Omega_A = \sum_{r=1}^{n} \nu_{Ar} J_r \ , \tag{3.27}$$

where J_r is defined as <u>the rate of the rth reaction</u>. It then follows that,

$$\sum_A \kappa_A \Omega_A = \sum_r \sum_A \kappa_A \nu_{Ar} J_r = \sum_r \mathcal{A}_r J_r \ , \tag{3.28}$$

where $\mathcal{A}_r := \sum_A \kappa_A \nu_{Ar}$ is the <u>affinity of the rth reaction</u> as defined in chapter 2. The advantages of using $J_r, \mathcal{A}_r$ as variables are that the J_r can be taken to be <u>independent</u> quantities, whereas the Ω_A must satisfy equation (3.11) and that in equilibrium $\mathcal{A}_r = 0$ for all r as was pointed out in chapter 2. With these definitions, (3.23) becomes,

$$* \quad \dot{s} + s\theta + \Big[\big(J_0{}^a - \sum_{A=1}^{N} \kappa_A J_A{}^a\big)/T\Big]_{;a} = \frac{1}{T}\Big\{\sum_{B=0}^{N} J_{Ba} X_B{}^a - \pi_{ab}\sigma^{ab} + \pi\theta - \sum_{r=1}^{n} \mathcal{A}_r J_r\Big\} \ . \tag{3.29}$$

Within the observer's frame we define,

$$* \qquad S^{*a} := su^a + (q^a - \sum_{A=1}^{N} K_A J_A)/T \; , \qquad (3.30)$$

so that (3.27) becomes, (in this frame)

$$* \qquad S^{*a}{}_{;a} = \frac{1}{T}\left\{ \sum_{B=0}^{N} J_{Ba} X_B{}^a - \pi_{ab}\sigma^{ab} + \pi\theta - \sum_{r=1}^{n} \mathcal{A}_{rr} J_r \right\} . \qquad (3.31)$$

In the observer's frame the first three components of S^{*a} give the usual entropy current density while the fourth gives the entropy density. The terms on the right hand side represent the entropy increase due to heat flow, diffusion, shear viscosity, bulk viscosity and chemical reactions respectively.

3.3 The Generalisation to Relativistic Thermodynamics.

At this stage it is necessary to decide on the generalisation from classical to relativistic thermodynamics. A very convenient assumption to make would be that S^{*a} transformed like a 4-vector, and was in fact equivalent to the entropy flux vector S^a of kinetic theory, as defined in the last chapter. It would then follow that s, T and K_A are scalars. However the entropy density measured in another frame would not have the value but would be,

$$\text{entropy density} = su^0 + (q^0 - \sum_{A=1}^{N} K_A J_A{}^0)/T \; .$$

Now it is clearly desirable that, at least in some cases, the entropy density should be a Lorentz invariant. Tolman (1934) has insisted that,

> "the entropy of a system should be unaltered by a reversible adiabatic change in velocity without absorbtion of heat."

He further remarks,

> "this requirement is in agreement with the statistical mechanical interpretation of entropy in terms of probability since the probability of finding a system in a given state should evidently be

independent of the velocity of the observer relative to it."

Now the definition of heat flow is somewhat arbitrary in relativistic thermodynamics. The energy flux w^a can be defined in a unique way as,

$$w^a := -T^{ab}u_b = \mu u^a + q^a , \qquad (q^a u_a = 0).$$

but the description of q^a as the heat flux and μu^a as the mass flux is slightly misleading since heat and energy are equivalent to mass in relativity theory. But however a heat flow is defined it will not vanish unless $q^a=0$, and in this case, (ignoring diffusion effects,) our entropy density reduces to the form,

$$S^{*a} = sN^a ,. \tag{3.32}$$

Tolman chooses a different way to generalise the classical theory. He extrapolates his requirement about the invariance of the entropy under an adiabatic change of reference frames to read,

> "the entropy density s is a Lorentz invariant quantity for both adiabatic and non-adiabatic changes."

It immediately follows that in Tolman's theory the entropy flux 4-vector has the form,

$$S_T{}^a = sN^a . \tag{3.33}$$

This assumption has been widely accepted,(see eg.Møller (1967) and works cited there,) and in such a theory it is necessary to assume that T transforms like a component of a 4-vector in order that a covariant form for the second law of thermodynamics can be established. This has led to a heated discussion as to whether a heated body appears cooler when in relative motion. It is clear that within the framework of thermodynamics there is no reason for preferring S^{*a} to $S_T{}^a$ or any other reasonable definition, as a choice for the entropy flux 4-vector. However if thermodynamics and kinetic theory are to be unified, the entropy flux definitions in both theories must be equivalent.

The kinetic theory definition,

$$S^a = -k \sum_A \int p^a(f_A \log f_A \mp \hat{f}_A \log \hat{f}_A)\pi_A \quad , \tag{2.29}$$

certainly reduces to both S^{*a} and $S_T{}^a$ for an equilibrium distribution function, but it is clearly unreasonable to expect S^a to have the special form (3.33) for arbitrary non-equilibrium distributions. Therefore, since we want to establish a link with kinetic theory, we shall reject Tolman's choice, and will generalise classical thermodynamics to the relativistic case by making the assumption,

" S^{*a} defined by equation (3.30) is the entropy flux 4-vector." (3.34)

The *'s before equations can now be dropped, and S^{*a} can be written as S^a without risk of ambiguity.

The covariant form of the second law of thermodynamics is now,

$$S^a{}_{;a} \quad 0 \quad , \tag{3.35}$$

which follows from the H-theorem of kinetic theory, (proved in the last chapter). We shall make the assumption that for small deviations from equilibrium, the quantities $J_B{}^a$,$(B=0,1,\ldots,N)$, ^{ab} , , J_r,$(r=1,2,\ldots,n)$, depend only linearly on the gradients of the equilibrium quantities, the $X_B{}^a$, $u_{(a;b)}$, $_r$. Then on comparing equations (3.29), (3.35) the following relations may be deduced:

$$J_A{}^a = \sum_{B=0}^{N} (L_{AB}u^a u^b + M_{AB}g^{ab})X_{Bb} \quad , \tag{3.36}$$

$$\pi^{ab} = -2\eta\sigma^{ab} \quad , \tag{3.37}$$

$$\pi = \zeta\theta + \sum_{r=1}^{n} \Lambda_r \mathcal{A}_r/T \quad , \tag{3.38}$$

$$J_r = \Lambda'_r \theta/T + \sum_{s=1}^{n} \Lambda_{rs}\mathcal{A}_s \quad , \quad r=1,2,\ldots,n, \tag{3.39}$$

where the scalar functions of state $L_{AB}, M_{AB}, \eta, \varsigma, \Lambda_r, \Lambda'_r, \Lambda_{rs}$ satisfy certain conditions. From (3.7), $J_{Aa}u^a = 0$, and since $X_{Aa}u^a \neq 0$ in general, we must have $L_{AB} = M_{AB}$, ie.,

$$J_A{}^a = \sum_{B=0}^{N} L_{AB}h^{ab}X_{Bb}. \tag{3.40}$$

Thus in the observer's frame, (3.40) reduces to its non-relativistic counterpart. From the usual Onsager-Casimir relations we know that is a symmetric matrix, and $\sum_A \sum_B L_{AB}h^{ab}X_{Aa}X_{Bb}$ is a positive semi-definite form, (assuming $T > 0$). Therefore the Onsager-Casimir relations hold also in the relativistic case, and in all reference frames,

$$L_{AB} = L_{BA}. \tag{3.41}$$

From (3.35), (3.37) it can be deduced that $\eta > 0$ and (3.38), (3.39) lead to,

$$\varsigma > 0 ,$$

$$\Lambda_r = \Lambda'_r ,$$

$$\Lambda_{rs} = \Lambda_{sr} ,$$

the other less well known Onsager-Casimir relations. (3.37)-(3.40) together with the Onsager-Casimir relations constitute the linear transport equations of relativistic non-equilibrium thermodynamics.

CHAPTER 4. SOLUTION OF THE BOLTZMANN EQUATION

4.1 Introduction and Discussion of Available Methods.

Before discussing the methods of obtaining approximate solutions of the Boltzmann equation one should be sure that the solutions sought do in fact exist. A somewhat restricted existence theorem has been given by Bichteler (1967) for a 1-component non-quantum gas. This result which is easily extended to a multicomponent or quantum gas guarantees the existence of a solution at least within a subregion of the domain of dependence, if the distribution function is required to be "nearly Maxwellian" on a certain Cauchy surface in phase space. A more precise statement and proof of the theorem can be found in Bichteler (1967).

Little progress has been made in constructing exact solutions of the Boltzmann equation, and so we shall review some of the techniques available for obtaining approximate solutions, assuming that the corresponding exact solutions do exist. All of the methods are purely formal, and no detailed analysis of the errors has been made.

1) Model Equations

The non-linear collision term is replaced by a simpler expression which preserves most of the physical properties. The best known of these is the BGK model of Bhatnagar, Gross and Krook (1954),

$$L(f) = C\,[f] = -(f - f_0)/\tau \;, \tag{4.1}$$

where f_0 is a local equilibrium distribution, (ie. one of the form (2.34) or (2.37) in which the α, β_a are functions of x,) and the interpretation of (4.1) is that collisions cause f to "relax" towards f_0 on a characteristic time scale τ . Clearly $1, p^a$ have to be collisional invariants, and this fixés the parameters describing f_0 but since these parameters are functionals of f there has been no real simplification. There is a numerical error in the values of the coefficients of viscosity and thermal conductivity when calculated by this method, and the approximation does not include the effects of thermal diffusion. On the other hand the form is particularly suited to numerical computation. For a relativistic calculation, see eg. Marle (1969).

2) Chapman-Enskog-Hilbert Expansions

This method was first developed by Hilbert (1912). More convenient forms were given by Chapman (1916), Enskog (1917), and a full account of the method can be found in Chapman & Cowling (1970).

The 1-component Boltzmann equation is the special case of,

$$L(f) = Q(f,f)/\varepsilon \quad , \tag{4.2}$$

where $Q(f,f)/\varepsilon$ is the usual quadratic collision integral when $\varepsilon = 1$. A solution of (4.2) is sought which is of the form,

$$f = f_0 + \varepsilon f_1 + \varepsilon^2 f_2 + \dots, \tag{4.3}$$

and which takes on initial values independent of ε . The method proceeds by formal substitution, and equating the coefficients of various powers of ε which acts merely as a "book-keeping" parameter. To zero order in ε

$Q(f_0,f_0) = 0$, and so f_0 is a locally Maxwellian or Fermi-Dirac (Bose-Einstein) distribution. The next order equation is a Fredholm integral equation for f_1 and the integrability conditions guarantee the validity of the Euler fluid equations to zeroth order. By successive iteration the fluid equations can be built up to any required degree of accuracy.

A defect of the method is that the solutions (4.3) are only regular in a neighbourhood of $\varepsilon = 0$, while the equation (4.2) is singular at $\varepsilon = 0$. In more physical terms, f_0 only depends on the five quantities, ρ, u^a, (normalised so that $u_a u^a = -1$), and T describing the fluid state, and all correction terms depend only on f_0 so that at each stage the solution depends only on the five fluid quantities. Such solutions are called normal solutions, and they are clearly only relevant to slight departures from equilibrium. Grad (1963) has shown that under certain conditions, an arbitrary distribution rapidly decays to a normal solution, and that the Hilbert scheme gives an asymptotic solution for large values of the time variable. Israel (1963) has explored the Chapman-Enskog method in the relativistic case, but his scheme contains certain difficulties, and some other methods of solving the integral equations have not yet produced useful results, (R K Sachs, private communication). In this context the next method might be useful. Marle (1969) has also studied the Chapman-Enskog method - in the relativistic case.

3) Variational Methods

These have been studied extensively in plasma kinetics, (eg. Robinson Bernstein (1952),) and are best suited to a linearised equation. For example if we write the first two terms in a Chapman-Enskog expansion as,

$$f = f_0(1 + \varepsilon\varphi) + O(\varepsilon^2),$$

then has to satisfy,

$$L(f_0) = -\tfrac{1}{2}\iiint W(p,p' \to p'',p''') f_0 f_0' (\varphi + \varphi' - \varphi'' - \varphi''')\pi'\pi''\pi''' ,$$

which we write as,

$$\psi = K\varphi \tag{4.4}$$

where $\psi = L(f_0)$ and K is a linear integral operator. We can define an inner product on the space of continuous bounded square integrable functions on momentum space by

$$(\varphi,\chi) := \int \varphi\chi\pi .$$

Then it is easily shown that

$$(\varphi,K\chi) += (\chi,K\varphi), \quad (\chi,K\chi) \leqslant 0,$$

* See also W A van Leeuwen (PhD thesis, Amsterdam, 1971).

with equality if and only if φ is a collisional invariant. It is also easily verified from the H-theorem that,

$$S^a{}_{;a} = -k(\varphi, K\varphi) \geqslant 0.$$

From thermodynamic considerations it might be expected that for the actual solution of the Boltzmann equation, $S^a{}_{;a}$ is a maximum compared with the values it would take for other possible distributions, and so by minimising $(\varphi, K\varphi)$ we approximate most closely to the actual solution. Indeed by considering the functional,

$$\lambda(\chi) = -(\psi, \chi)^2/(\chi, K\chi),$$

and varying χ one finds that λ is stationary if and only if χ is proportional to a solution of (4.4). Thus trial solutions arbitrarily close to an exact solution can be found. However one usually wants to evaluate transport coefficients and a variational problem can be posed just as in the Rayleigh-Ritz method so that the maximum to be found gives the value of the transport coefficient. This method has the advantage that if the trial function differs from the actual one by ε, the error in the transport coefficient is $O(\varepsilon^2)$, so that the method is potentially an accurate one.

4) The Method of Moments

The basic idea of the method of moments is very simple. In most kinetic theory problems we are not so much interested in the distribution function as in its first few moments. If we multiply the Boltzmann equation by $1, p^a, p^a p^b, \ldots,$ successively and integrate over phase space then the left hand sides can be transformed into expressions involving the partial derivatives of the moments. If an approximation procedure can be devised for evaluating the collision integrals which occur on the right hand sides of the equations, then we shall have obtained a set of partial differential equations for the moments of the distribution function. This method was first suggested by Maxwell (1867) but its full power was first exploited by Grad (1949) (1958) and refined by Mintzner (1965). A recent comprehensive survey of the classical theory has been given by Burgers (1969). Relativistic treatments have been given by Chernikov(1963), Stewart & Anderson, (Stewart(1969), Anderson(1970) and Marle (1966), (1969). Chernikov's method suffers from certain difficulties which will be discussed later. Marle's method is very similar to the one described here, but Marle restricts his discussion to a 1-component gas whose particles have non-zero rest mass, and to the BGK approximation for the collision integral. In this work we shall study a multicomponent gas, whose particles can have arbitrary (fixed) rest masses. Both elastic and inelastic collisions as well as creation and annihilation processes will be permitted, and the Boltzmann collision integral will be used. Although the presentation will mainly be classical (non-quantum), a section will discuss the inclusion of quantum effects. An electromagnetic field has not been included in the discussion.* It is a trivial matter to include a background electromagnetic field, but there are serious convergence problems if the particles act as the sources of the field.

4.2 The Moment Equations

The non-quantum Boltzmann equation was derived in chapter 2 in the form (2.20),

$$L_A(f_A) = \tfrac{1}{2}\sum_{B,C,D}\iiint(f_C f_D - f_A f_B)W_{AB\to CD}\,\pi_B\pi_C\pi_D +$$

$$+\tfrac{1}{2}\sum_{B,C}\iint(f_B f_C - f_A)W_{BC\to A}\,\pi_B\pi_C - \sum_{B,C}\iint(f_A f_B - f_C)W_{AB\to C}\,\pi_B\pi_C\,, \tag{4.5}$$

for the A-component in a multicomponent gas. The three terms on the right hand side represent the rate of increase of f_A due to binary collisions, creation, and annihilation processes respectively, while the left hand side gives the convective rate of change of f_A in phase space,

*See also Israel (preprint, 1971)

$$L_A(f_A) = p_A{}^a \partial/\partial x^a - \Gamma^a{}_{bc} p_A{}^b p_A{}^c \partial/\partial p_A{}^a \quad ,$$

(electromagnetic effects are being ignored). Equation (4.5) also applies to the 1-component case if the creation and annihilation terms are dropped, the summation is omitted, and the factor $\frac{1}{2}$ incorporated into W giving,

$$L(f) = \iiint (f''f''' - ff')W(p,p' \rightarrow p'',p''')\pi'\pi''\pi''' \quad . \tag{4.6}$$

If $\overset{n}{g}_A{}^{\underset{\sim}{a}}$ represents a product of n p_A's eg. $1, p_A{}^a, p_A{}^a p_A{}^b, \ldots$ then it is easy to verify, either by direct differentiation or from the lemma in chapter 2, that,

$$\int_{P_A(x)} \overset{n}{g}_A{}^{\underset{\sim}{a}} L_A(f_A)\pi_A = \left\{ \int_{P_A(x)} \overset{n}{g}_A{}^{\underset{\sim}{a}} p_A{}^c f_A(x,p_A)\pi_A \right\}_{;c} \quad . \tag{4.7}$$

In particular taking $\overset{n}{g}_A{}^{\underset{\sim}{a}} = 1, p_A{}^a, \ p_A{}^a p_A{}^b,$ gives,

$$\int L_A(f_A)\pi_A = N_A{}^a{}_{;a} \quad , \tag{4.8}$$

$$\int p_A{}^a L_A(f_A)\pi_A = T_A^{ab}{}_{;b} , \tag{4.9}$$

$$\int p_A^a \, p_A^b \, L_A(f_A)\pi_A = S_A{}^{abc}{}_{;c} \quad , \tag{4.10}$$

where the notation for the moments introduced in chapter 2 has been used. It should be noted that equations (4.8)-(4.10) are exact and do not depend on the details of the collision processes. The derivatives on the right hand sides express purely kinematical effects.

If we now consider the right hand sides of the Boltzmann equation we obtain,

$$\int g_A{}^{\underset{\sim}{a}} L_A(f_A)\pi_A = \tfrac{1}{2}\sum_{B,C,D} \iiiint \overset{n}{g}_A{}^{\underset{\sim}{a}}(f_C f_D - f_A f_B)W_{ABCD}\pi^4 \; +$$

$$+ \tfrac{1}{2}\sum_{B,C} \iiint \overset{n}{g}_A{}^{\underset{\sim}{a}}(f_B f_C - f_A)W_{BCA}\pi^3 - \sum_{B,C} \iiint \overset{n}{g}_A{}^{\underset{\sim}{a}}(f_A f_B - f_C)W_{ABC}\pi^3 \quad , \tag{4.11}$$

where $W_{AB\to CD}\pi_A\pi_B\pi_C\pi_D$ has been abbreviated to $W_{ABCD}\pi^4$ etc. In particular if $n = 0$,

$$\Omega_A = N_A{}^a{}_{;a} = \tfrac{1}{2}\sum_{B,C,D}\iiiint(f_Cf_D-f_Af_B)W_{ABCD}\pi^4 + \tfrac{1}{2}\sum_{B,C}\iiint(f_Bf_C-f_A)W_{BCA}\pi^3 -$$

$$- \sum_{B,C}\iiint(f_Af_B-f_C)W_{ABC}\pi^3 \quad , \tag{4.12}$$

is the rate of increase of A-particles due to chemical reactions. If only elastic collisions occur, then a little manipulation of equation (4.12) gives

$$\Omega_A = N_A{}^a{}_{;a} = 0 \ ,$$

which shows that the number densities of individual components are conserved.

If equation (4.11) is summed over all components, a little manipulation gives,

$$\sum_A\int \overset{n}{g}_A{}^{\underset{\sim}{a}} L_A(f_A)\pi_A = 1/8\sum_{A,B,C,D}\iiiint(\overset{n}{g}_A{}^{\underset{\sim}{a}}+\overset{n}{g}_B{}^{\underset{\sim}{a}}-\overset{n}{g}_C{}^{\underset{\sim}{a}}-\overset{n}{g}_D{}^{\underset{\sim}{a}})(f_Cf_D-f_Af_B)W\pi^4 \ +$$

$$+ \tfrac{1}{2}\sum_{A,B,C}\iiint(\overset{n}{g}_C{}^{\underset{\sim}{a}}-\overset{n}{g}_A{}^{\underset{\sim}{a}}-\overset{n}{g}_B{}^{\underset{\sim}{a}})(f_Af_B-f_C)W_{ABC}\pi^3 \ , \tag{4.13}$$

where the reversibility of collisions has been used. If we set $n = 0$ in equation (4.13) we obtain,

$$N^a{}_{;a} = -\tfrac{1}{2}\sum_{A,B,C}\iiint(f_Af_B-f_C)W_{ABC}\pi^3,$$

and if only binary collisions occur,

$$N^a{}_{;a} = 0,$$

the usual number conservation law. Since 4-momentum is conserved in collisions, setting $n = 1$ in equation (4.13) gives,

$$T^{ab}{}_{;b} = 0,$$

the usual conservation law. Because this last equation can be regarded as an integrability equation for the Einstein field equations we see that in general the field equations can only be integrated if momentum is conserved in all microscopic collisions.

We can immediately deduce from equation (4.11) the partial differential equations satisfied by the moments when equilibrium occurs, ie. when there are no collisions, or when the effects of collisions cancel out, (detailed balancing). They are,

$$N_A{}^a{}_{;a} = 0,$$

$$T_A{}^{ab}{}_{;b} = 0,$$

$$S_A{}^{abc}{}_{;c} = 0, \text{ etc.}$$

In the next three sections we shall discuss how to construct analogous, but more complicated equations in the case where the collision term does not vanish. However these new equations will be approximations only, while the previous equations have all been exact.

4.3 The Approximation of the Distribution Function

If the collision term in equation (4.11) does not vanish, we would like to approximate it in some way, for in general exact solutions cannot be hoped for. If we consider only small deviations from equilibrium then $f(x,p)$ will be nearly an equilibrium distribution $f_0(x,p)$ and we could write,

$$f(x,\bar{p}) \approx f_0(x,p)\left\{ 1 + \sum_{n=1}^{m} \overset{n}{a}_{a_1 a_2 \ldots a_n} p^{a_1} p^{a_2} \ldots p^{a_n} \right\} , \qquad (4.14)$$

where for the moment the particle species index has been dropped. A similar approximation could be made in situations far from equilibrium, if a suitable (not necessarily equilibrium) f_0 was known. f_0 is assumed strictly positive and such that all the moments $\int \overset{n}{g}\, \underset{\sim}{a} f_0 \pi$ exist. We would then choose the $\overset{n}{a}_{\underset{\sim}{a}}$ for $1 \leq n \leq m$ in such a way as to give the most accurate representation for $f(x,p)$ and substitute this expression for $f(x,p)$ into the collision integrals. These integrals can be explicitly evaluated since they only involve f_0 and polynomials in p^a. By multiplying (4.14) by each ${}^n_{\,g}{}^{\underset{\sim}{a}}$ in turn and integrating we see that the ${}^n a_{\underset{\sim}{a}}$ are linear combinations of the moments of $f(x,p)$. Since the collision integrals are quadratic in f, this approximation will produce a quadratic form in the moments of f, and so equations (4.11) can be approximated by a set of partial differential equations in the moments of f. However the ${}^n a_{\underset{\sim}{a}}$ depend crucially on m, and the expressions for the ${}^n a_{\underset{\sim}{a}}$ are rather complicated. There are also some serious convergence difficulties. For these reasons we adopt a slightly more sophisticated approach.

We first demonstrate the existence of a set of polynomials $H_n{}^{\underset{\sim}{a}}(x,p)$ of degree n in p^a which are orthogonal with $f_0(x,p)$ as weight function, ie.,

$$(H_n{}^{a},H_m{}^{b}) := \int_{P_m(x)} H_n{}^{\underset{\sim}{a}} H_m{}^{\underset{\sim}{b}} f_0(x,p)\pi = \delta_{mn} M_n{}^{\underset{\sim}{a}\underset{\sim}{b}}(x) , \qquad (4.15)$$

(there is no summation convention for the indices $m,n,\ldots$ describing the order of the polynomials. Any intended summation will be shown explicitly.) In the classical case this is easy, (see eg. Grad (1949),) for the required polynomials are the Hermite polynomials, and these also satisfy a Rodrigues relation. Unfortunately there is no set of orthogonal polynomials which also satisfies a Rodrigues relation in the relativistic case. Chernikov (1963) chose to generalise the classical treatment by retaining the Rodrigues relation, and so his polynomials are not orthogonal. They therefore suffer from all of the defects mentioned at the end of the last paragraph. We shall not adopt this approach and instead we construct the polynomials from the set $\{{}^n_{\,g}{}^{\underset{\sim}{a}}\}$ by essentially Gram-Schmidt orthogonalisation. A similar, but more rigorous treatment has been given by Marle (1966),(1969). We can obviously take

$$H_0 = 1, \qquad (4.16)$$

so that,

$$M_0 = A_0(x), \tag{4.17}$$

(The notation for the moments is given in chapter 2, equations (2.21)-(2.26). When the suffices in tensors such as $H_n{}^a$, $M_n{}^{ab}$, etc. are explicitly written out, we shall omit the suffix n, eg. H^{ab}, M^{abcd} instead of $H_2{}^{ab}$, $M_2{}^{abcd}$.) We write H^a in the form

$$H^a = p^a - \alpha^a(x), \tag{4.18}$$

where α^a is to be determined. From the orthogonality condition (4.15) we have,

$$0 = (H, H^a) = N_0{}^a - \alpha^a A_0,$$

and so,

$$\alpha^a = N_0{}^a / A_0. \tag{4.19}$$

Further,

$$M^{ab} = (H^a, H^b) = T_0{}^{ab} - \alpha^a \alpha^b A_0. \tag{4.20}$$

We write H^{ab} in the form,

$$H^{ab} = p^a p^b - \alpha^{ab}{}_c H^c - \beta^{ab}. \tag{4.21}$$

The orthogonality property (4.15) implies,

$$0 = (H^{ab}, H) = T_0{}^{ab} - \beta^{ab}/A_0,$$

and so,

$$\beta^{ab} = T_0{}^{ab}/A_0. \tag{4.22}$$

Also,

$$0 = (H^{ab}, H^c) = S_0{}^{abc} - \alpha^{ab}{}_d M^{cd} - T_0{}^{ab} \alpha^c,$$

and so,

$$M^{cd} \alpha^{ab}{}_d = S_0{}^{abc} - T_0{}^{ab} \alpha^c. \tag{4.23}$$

Further,

$$M^{abcd} = (H^{ab}, H^{cd}) = Q_0{}^{abcd} - \alpha^{ab}{}_e \alpha^{cd}{}_f M^{ef} - \beta^{ab} \beta^{cd} A_0. \tag{4.24}$$

Finally we shall need,

$$H^{abc} = p^a p^b p^c - \alpha^{abc}{}_{de} H^{de} - \beta^{abc}{}_d H^d - \gamma^{abc}, \qquad (4.25)$$

where

$$M^{defg} \alpha^{abc}{}_{fg} = P_0{}^{abcde} - \alpha^{de}{}_f Q_0{}^{abcf} - \beta^{de} S_0{}^{abc}, \qquad (4.26)$$

$$M^{de} \beta^{abc}{}_e = Q_0{}^{abcd} - \alpha^d S_0{}^{abc}, \qquad (4.27)$$

$$M_0 \gamma^{abc} = S_0{}^{abc}. \qquad (4.28)$$

Higher order polynomials can obviously be constructed in an analogous way. It is clear that the polynomials $H_r{}^{\underset{\sim}{a}}$ of a given order $r \gg 2$ are not all independent. Since the contraction of any polynomial of order r gives a polynomial of lower order (because $p_a p^a = -m^2$), such contractions must be orthogonal to the original $H_r{}^{\underset{\sim}{a}}$ and so the contraction is self orthogonal. Hence

$$H_2{}^a{}_a = 0,$$

with corresponding results for higher orders. Since the homogeneous systems corresponding to equations (4.23), (4.26), (4.27) etc. have non-trivial solutions, it is not immediately clear that the inhomogeneous equations are soluble. But if $q_{\underset{\sim}{a}} M_r{}^{\underset{\sim}{a}\underset{\sim}{b}} = 0$, then also, $q_{\underset{\sim}{a}} H_r{}^{\underset{\sim}{a}}$

and so $q_{\underset{\sim}{a}}$ is orthogonal to the right hand side of the inhomogeneous equation in question. Hence equations such as (4.23), (4.26), (4.27) etc. are soluble.

We now try to approximate f by an expansion,

$$f \approx f_0 \Big(\sum_{n=0}^{M} \overset{n}{a}_{\underset{\sim}{a}} H_n{}^{\underset{\sim}{a}} \Big). \qquad (4.29)$$

For the moment we assume that the right hand side converges to f as M

If we then multiply equation (4.29) by each $H_n^{\underset{\sim}{a}}$ in turn and integrate over the mass shell, (assuming sufficiently strong convergence properties,) we obtain expressions for the unknown coefficients $\overset{n}{a}_{\underset{\sim}{a}}$,

$$M_n^{ab}(x)a_b(x) = \int f(x,p)H_n^{\underset{\sim}{a}}(x,p)\pi , \qquad (4.30)$$

These equations can be solved for the $\overset{n}{a}_{\underset{\sim}{a}}$ because they are of the same form as (4.23), (4.26), (4.27) etc. which have already been shown to be soluble. The first few equations of the set (4.30) are,

$$A_0\overset{0}{a} = A \text{ , or } \overset{0}{a} = A/A_0 , \qquad (4.31)$$

$$M^{ab}a = N^a - \alpha^a A = N^a - N_0^a A/A_0 , \qquad (4.32)$$

$$M^{abcd}a_{cd} = T^{ab} - \alpha^{ab}{}_c(N^c - N_0^c A/A_0) - T_0^{ab}A/A_0 , \qquad (4.33)$$

$$M^{abcdef}a_{def} = S^{abc} - \alpha^{abc}{}_{de}\left[T^{de} - \alpha^{de}{}_f(N^f - N_0^f A/A_0) - T_0^{de}A/A_0\right] - \beta^{abc}{}_d(N^d - N_0^d A/A_0) - S_0^{abc}A/A_0 . \qquad (4.34)$$

Thus each $\overset{n}{a}_{\underset{\sim}{a}}$ can be determined, (except for its contractions,) in a form which is independent of M . This is the advantage of using the slightly more complicated expansion (4.29) instead of (4.14).

The expression (4.29) will converge sufficiently strongly for operations such as term by term integration to be valid if $\int f^2/f_0 \, \pi$ is finite, and we shall always assume that this criterion is satisfied. It then follows that the expression (4.29) converges to f in the mean, because the $H_n^{\underset{\sim}{a}}$ obviously form a complete set. (For a proof see e.g. Marle (1969).)

To see what it means physically we first consider a shockwave in which the far upstream and far downstream distributions are approximately Maxwellian with temperatures T_u , T_d . If we try to approximate the downstream distribution f_d in terms of a Maxwellian distribution with temperature T_u the convergence criterion will only be satisfied for weak shocks with Mach number $M \lesssim 1.8$ since we must have $T_d < 2T_u$. However the upstream distribution can always be expanded in terms of a Maxwellian distribution with temperature T_d

for arbitrarily strong shocks. To see why, consider two 1-dimensional approximately Maxwellian distributions with temperatures T_d, T_u, $(T_d < T_u)$. The distribution function at the higher temperature has a flatter peak and higher tails than the other. Moments of low order will be of approximately the same magnitude for both distributions, but for high orders, where the main contribution is from the tails, the moments of the high temperature distribution will be much greater than those of the low temperature one. Thus if, in the expansion (4.29) f_0 has a much lower "temperature" than f the $\overset{n}{a}_{\underset{\sim}{a}}$ which are of the order of magnitude of the nth moments, will be large, and the series will not converge as $M \to \infty$. Conversely if f_0 has a "higher" "temperature" than f the $\overset{n}{a}_{\underset{\sim}{a}} \to 0$ sufficiently strongly for the required convergence properties.

4.4 Matching Conditions

Although it is possible to develop the theory using equations (4.31)-(4.34) as they stand, considerable simplifications can be made exploiting the arbitrariness in $f_0(x,p)$. If $f_0(x,p)$ is a locally Maxwellian distribution it contains five arbitrary parameters, (essentially a specification of n_0 the number density, $u_0^{\ a}$ the normalised 4-velocity, and T_0 the temperature), and we shall assume that there are five such adjustable parameters for more general choices of f_0. The specification of these parameters involves two problems, the first of which occurs even in the classical theory. It is difficult to specify a mean 4-velocity in a non-equilibrium fluid, so as to use up three of the five degrees of freedom, in a meaningful way. A four velocity is the normalised flux of some quantity, and in general, different fluxes have different directions. Of course in equilibrium they are all parallel, and for small deviations from equilibrium they are nearly parallel. In this special case we can specify a four velocity $u_0^{\ a}$ which partially fixes f_0, and which has some physical significance. A "density" and "temperature" can also be specified. However in the general case it is not possible to specify these quantities in a meaningful way.

The second difficulty arises from the arbitrariness in the choice of u_0^a. The natural question to ask is whether the method is Lorentz invariant; are the sets of transport equations which result from different choices of u_0^a connected by the appropriate Lorentz transformation? In section 4.8 we shall show that the method is indeed Lorentz invariant for small deviations from equilibrium.

We first consider the matching conditions for a 1-component gas. Inspection of equations (4.31),(4.32) shows that considerable simplification occurs if we take,

$$A_0 = A \quad , \qquad N_0^a = N^a, \qquad (4.35)$$

to specify $f_0(x,p)$. In the notation of chapter 3 condition (4.35) requires,

$$n_0 = n, \quad u_0^a = u^a, \quad \mu_0 - 3p_0 = \mu - 3p. \qquad (4.36)$$

This is not the only choice. Eckart (1940) has suggested as a matching condition,

$$n_0 = n, \quad u_0^a = u^a \ , \quad \mu_0 = \mu \ ,$$

and Landau & Lifschitz (1959) have used,

$$\mu_0 u_0^a = \mu u^a - q^a, \quad n_0 = n \ .$$

The effect of these will be discussed later but it is obvious that in this context, the matching conditions (4.35) should be used. Then equations (4.31)-(4.34) simplify to give

$$\dot{a} = 1, \qquad (4.31')$$

$$a_a = 0 \qquad (4.32')$$

$$M^{abcd} a_{cd} = \Delta T^{ab} := T^{ab} - T_0^{ab} \ , \qquad (4.33')$$

$$M^{abcdef} a_{def} = S^{abc} - S_0^{abc} - \alpha^{abc}{}_{de} \Delta T^{de} \ . \qquad (4.34')$$

Although section 4.3 was written in the notation of a 1-component fluid, it is clear that the considerations invoked there apply equally well to each component of a multicomponent fluid. Again we could choose a matching condition (4.35) and obtain a set of equations (4.31')-(4.34') for each component of the fluid. However it would be more convenient to describe diffusion in terms of the matching conditions,

$$A_{Ao} = A_A \ , \ u_{Ao}{}^a = u^a, \quad (N_{Ao}{}^a - N_A{}^a)u_a = 0, \ A=1,2,\ldots,N, \tag{4.37}$$

where u^a is a mean 4-velocity for the gas as a whole. In the notation of chapter 2 we have the conditions,

$$\mu_A - 3p_A = \mu_{Ao} - 3p_{Ao}, \quad u_{Ao}{}^a = u^a, \quad n_{Ao} = n_A, \quad A=1,2,\ldots,N. \tag{4.38}$$

Then the diffusion flux is,

$$J_A{}^a = h^{ab} N_{Ab} = N_A{}^a - N_{Ao}{}^a \ , \tag{4.39}$$

and the matching conditions simplify to give,

$$a = 1, \tag{4.31''}$$

$$M_A{}^{ab} a_{Ab} = J_A{}^a, \tag{4.32''}$$

$$M_A{}^{abcd} a_{Acd} = \Delta T_A{}^{ab} - \alpha_A{}^{ab}{}_c J_A{}^c \ , \tag{4.33''}$$

$$M_A{}^{abcdef} a_{Adef} = S_A{}^{abc} - S_{Ao}{}^{abc} - \alpha_A{}^{abc}{}_{de}(\Delta T_A{}^{de} - \alpha_A{}^{de}{}_f J_A{}^f) - \beta_A{}^{abc}{}_d J_A^d, \tag{4.34''}$$

which differ from (4.31')-(4.34') only in the diffusion terms.

Although in this matching scheme each component can have an equilibrium distribution, the temperatures might be different for each component. In most astrophysical applications collisional relaxation times are much shorter than reaction times, and in this case it is more appropriate to give the f_{Ao}'s a common temperature. Then however, the chemical affinities will not be zero in the unperturbed f_{Ao}.) We shall therefore use the physically more relevant matching conditions,

$$A_{Ao} = A_A \ , \ u_{Ao}^a = u^a, \ T_{Ao} = T, (\text{or } \beta_{Ao} = \beta_A), \ A=1,2,\ldots,N, \tag{4.38'}$$

where u^a, T are some, (not further specified,) mean 4-velocity and temperature.

Then if we define ε_A by,

$$N_A^a - N_{Ao}^a = \varepsilon_A u^a + J_A^a , \qquad (4.39')$$

the matching conditions simplify to give,

$$a_A = 1, \qquad (4.31''')$$

$$M_A^{ab} a_{Ab} = \varepsilon_A u^a + J_A^a , \qquad (4.32''')$$

$$M_A^{abcd} a_{Adef} = \Delta T_A^{ab} - \alpha_A^{ab}{}_c (\varepsilon_A u^c + J_A^c), \qquad (4.33''')$$

$$M_A^{abcdef} \cdot a_{Adef} = S_A^{abc} - S_{Ao}^{abc} - \alpha_A^{abc}{}_{de}\left[\Delta T_A^{de} - \alpha_A^{de}{}_f (\varepsilon_A u^f + J_A^f)\right] - \beta_A^{abc}{}_d (\varepsilon_A u^d + J_A^d), \qquad (4.34''')$$

which differ from (4.31")- (4.34") only in the ε_A terms.

4.5 The Truncation & Linearisation Procedure

We shall now reconsider the basic moment equations (4.11) and replace each distribution function occurring on the right hand side by the appropriate expansion (4.29) to obtain,

$$\int \overset{n}{g}{}_A^a L_A(f_A)\, \pi_A = \sum_{r,s} \sum_{B,C,D} \left\{ {}^{nrs}C_{\underset{\sim}{ABCD}}^{\underset{\sim}{ade}}\, \overset{r}{a}_{C\underset{\sim}{d}}\, \overset{s}{a}_{D\underset{\sim}{e}} - {}^{nrs}D_{\underset{\sim}{ABCD}}^{\underset{\sim}{ade}}\, \overset{r}{a}_{A\underset{\sim}{d}}\, \overset{s}{a}_{B\underset{\sim}{e}} \right\} + \sum_{r,s} \sum_{B,C} \left\{ {}^{nrs}E_{\underset{\sim}{ABC}}^{\underset{\sim}{acd}}\, \overset{r}{a}_{B\underset{\sim}{c}}\, \overset{s}{a}_{C\underset{\sim}{d}} - {}^{nrs}F_{\underset{\sim}{ABC}}^{\underset{\sim}{acd}}\, \overset{r}{a}_{A\underset{\sim}{b}}\, \overset{s}{a}_{B\underset{\sim}{c}} \right\} + \sum_{r} \sum_{B,C} \left\{ {}^{nr}G_{\underset{\sim}{ABC}}^{\underset{\sim}{ad}}\, \overset{r}{a}_{C\underset{\sim}{d}} - {}^{nr}K_{\underset{\sim}{ABC}}^{\underset{\sim}{ad}}\, \overset{r}{a}_{A\underset{\sim}{d}} \right\} , \qquad (4.40)$$

where

$${}^{nrs}C_{\underset{\sim}{ABCD}}^{\underset{\sim}{ade}} := \tfrac{1}{2}\iiiint \overset{n}{g}{}_A^{\underset{\sim}{a}}\, H_{Cr}^{\underset{\sim}{d}}\, H_{Ds}^{\underset{\sim}{e}}\, f_{Co}\, f_{Do}\, W_{ABCD}\, \pi^4 , \qquad (4.41)$$

$${}^{nrs}D_{\underset{\sim}{ABCD}}^{\underset{\sim}{ade}} := \tfrac{1}{2}\iiiint \overset{n}{g}{}_A^{\underset{\sim}{a}}\, H_{Ar}^{\underset{\sim}{d}}\, H_{Bs}^{\underset{\sim}{e}}\, f_{Ao}\, f_{Bo}\, W_{ABCD}\, \pi^4 , \qquad (4.42)$$

$$ {}^{nrs}E^{\underset{\sim}{a}\underset{\sim}{c}\underset{\sim}{d}}_{ABC} := \frac{1}{2}\iiint \overset{n}{g}_A{}^{\underset{\sim}{a}} H^{\underset{\sim}{c}}_{Br} H^{\underset{\sim}{d}}_{Cs} \, f_{Bo} f_{Co} W_{BCA} \pi^3 \quad , \tag{4.43} $$

$$ {}^{nrs}F^{\underset{\sim}{a}\underset{\sim}{c}\underset{\sim}{d}}_{ABC} := \iiint \overset{n}{g}_A{}^{\underset{\sim}{a}} H^{\underset{\sim}{c}}_{Ar} H^{\underset{\sim}{d}}_{Bs} \, f_{Ao} f_{Bo} W_{ABC} \pi^3 \quad , \tag{4.44} $$

$$ {}^{nr}G^{\underset{\sim}{a}\underset{\sim}{d}}_{ABC} := \iiint \overset{n}{g}_A{}^{\underset{\sim}{a}} H^{\underset{\sim}{d}}_{Cr} f_{Co} W_{ABC} \pi^3 \, , \tag{4.45} $$

$$ {}^{nr}K^{\underset{\sim}{a}\underset{\sim}{d}}_{ABC} := \frac{1}{2}\iiint \overset{n}{g}_A{}^{\underset{\sim}{a}} H^{\underset{\sim}{a}}_{Ar} f_{Ao} W_{BCA} \pi^3 \, , \tag{4.46} $$

and all of the collision integrals can be explicitly evaluated since they involve f_{Ao} rather than f_A . The nth equation in the set (4.40) expresses the divergence of the (n+1)th moment on the left hand side in terms of a_A's on the right hand side.

From the previous section it is obvious that the right hand side of the nth equation in the set (4.40) involves moments of all orders. For practical purposes it is necessary to truncate this set in some way so as to obtain a usable set of equations. The most convenient way of doing this is to note that if the expansion (4.29) converges as $M \to \infty$ then,

$$ \| H^{\underset{\sim}{a}}_{n} \| \cdot | \overset{n}{a}_{A\underset{\sim}{a}} | \longrightarrow 0 \quad \text{as } n \to \infty \quad , $$

so that a good approximation might be given by

$$ \overset{n}{a}_{A\underset{\sim}{a}} = 0 \quad \text{for } n \;\; n_0 \;, \quad A=1,2,\ldots,N. \tag{4.47} $$

In this case the right hand sides of the equations in the set (4.40) involve moments of up to the n_0th order only. The left hand sides of the first (n_0-1) equations would involve moments of up to the n_0th order, but the n_0th equation would involve the divergence of the (n_0+1)th order. However the equation,

$$\overset{n_o+1}{a}_{A\underset{\sim}{a}} = 0,$$

expresses the (n_o+1)th moment in terms of moments of lower order, and so the effect of the truncation procedure (4.47) is to produce a set of n_o first order partial differential equations for the first n_o moments of the distribution function for each species.

However a second approximation can be made which is also consistent with the requirement that $|\overset{n}{a}_{A\underset{\sim}{a}}| \rightarrow 0$ as $n \rightarrow \infty$. Clearly the $\overset{n}{a}_{A\underset{\sim}{a}}$ for $n \geqslant 1$ have to be "small", and so the right hand sides of the equations (4.40) which are bilinear in the $\overset{n}{a}_{A\underset{\sim}{a}}$ can be linearised. Using the matching condition $\overset{0}{a}_A = 1$ we can obtain,

$$\left[\int \overset{n}{\underset{\sim}{g}}{}_A{}^{\underset{\sim}{a}} p_A{}^c f_A \pi_A \right]_{;c} = \sum_r \sum_E {}^{nr}B^{\underset{\sim}{a}\underset{\sim}{b}}_{AE} \overset{r}{a}_{E\underset{\sim}{b}} \quad , E=A,B,C,D, \tag{4.48}$$

where

$$^{nr}B^{\underset{\sim}{a}\underset{\sim}{b}}_{AA} := -\sum_{B,C,D} {}^{nr0}D^{\underset{\sim}{a}\underset{\sim}{b}}_{ABCD} - \sum_{B,C}\left[{}^{nr0}F^{\underset{\sim}{a}\underset{\sim}{b}}_{ABC} + {}^{nr}K^{\underset{\sim}{a}\underset{\sim}{b}}_{ABC}\right], \tag{4.49}$$

$$^{nr}B^{\underset{\sim}{a}\underset{\sim}{b}}_{AB} := -\sum_{C,D} {}^{n0r}D^{\underset{\sim}{a}\underset{\sim}{b}}_{ABCD} + \sum_{C}\left[{}^{nr0}E^{\underset{\sim}{a}\underset{\sim}{b}}_{ABC} - {}^{n0r}F^{\underset{\sim}{a}\underset{\sim}{b}}_{ABC}\right] , \tag{4.50}$$

$$^{nr}B^{\underset{\sim}{a}\underset{\sim}{b}}_{AC} := \sum_{B,D} {}^{nr0}C^{\underset{\sim}{a}\underset{\sim}{b}}_{ABCD} + \sum_{B}\left[{}^{n0r}E^{\underset{\sim}{a}\underset{\sim}{b}}_{ABC} \quad {}^{nr}G^{\underset{\sim}{a}\underset{\sim}{b}}_{ABC}\right] , \tag{4.51}$$

$$^{nr}B^{\underset{\sim}{a}\underset{\sim}{b}}_{AD} := \sum_{B,C} {}^{n0r}C^{\underset{\sim}{a}\underset{\sim}{b}}_{ABCD} . \tag{4.52}$$

Thus the truncation and linearisation conditions lead to a set of n_o approximate first order linear partial differential equations for the first n_o moments of f and these are the required moment equations.

In the subsequent work we shall take $n_o = 2$, since this approximation will give equations for the second moment of f_A , and in particular it will lead to expressions for the heat flow, viscous stresses and diffusion fluxes from which we can deduce transport equations. In this respect the relativistic theory is simpler than the classical one, for in the latter case the heat flow involves a third order moment, and we should have to set $n_0'=3$. It may be that the accuracy of our set of equations (4.40) or (4.48) improves as n_0 is increased, but this aspect has not been considered here. The best practical check on such approximations is to see if they agree with experiment. Satisfactory agreement has been established in the classical case for the 13-moment approximation of Grad, which corresponds closely to the $n_0'=2$ case presented here. In principle all the higher order moments can be calculated from the lower order ones once the truncation approximation (4.47) has been adopted. It is then possible to check whether these moments satisfy the appropriate higher order equations of the set (4.40). This could also be used to check the validity of the approximation.

In the 1-component case the equation (4.48) simplify considerably, but the required equations can be obtained more directly. We multiply the 1-component Boltzmann equation, (4.6),

$$L(f) = \iiint (f''f'''-ff')W(p,p'\rightarrow p'',p''')\pi'\pi''\pi''' ,$$

by each $\overset{n}{g}{}^{a}$ in turn, and integrate to obtain,

$$\left\{\int \overset{n}{g}{}^{\underset{\sim}{a}} p^c f\, \pi\right\}_{;c} = \iiiint \overset{n}{g}{}^{\underset{\sim}{a}}(f''f''' - ff')W\pi^4 .$$

Using the symmetry and reversibility properties of collisions we have,

$$\begin{aligned}\left[\int \overset{n}{g}{}^{\underset{\sim}{a}} p^c f\, \pi\right]_{;c} &= \tfrac{1}{2}\iiiint (\overset{n}{g}{}^{\underset{\sim}{a}}+\overset{n}{g}{}'^{\underset{\sim}{a}})(f''f''' - ff')W\pi^4 ,\\ &= \tfrac{1}{4}\iiiint (\overset{n}{g}{}^{\underset{\sim}{a}}+\overset{n}{g}{}'^{\underset{\sim}{a}}-\overset{n}{g}{}''^{\underset{\sim}{a}}-\overset{n}{g}{}'''^{\underset{\sim}{a}})(f''f''' - ff')W\pi^4 ,\\ &= -\tfrac{1}{2}\iiiint (\overset{n}{g}{}^{\underset{\sim}{a}}+\overset{n}{g}{}'^{\underset{\sim}{a}}-\overset{n}{g}{}''^{\underset{\sim}{a}}-\overset{n}{g}{}'''^{\underset{\sim}{a}})(ff')W\pi^4 ,\\ &= \iiiint (\overset{n}{g}{}''^{\underset{\sim}{a}}-\overset{n}{g}{}^{\underset{\sim}{a}})ff'W\pi^4 . \end{aligned} \tag{4.53}$$

We now substitute the expansion (4.29) for each f on the right hand side, truncate, and linearise to obtain,

$$\left[\int g^{\underset{\sim}{a}}\, p^c f\right]_{;c} = 2\sum_{r=0}^{2} {}^{nr}B^{\underset{\sim}{a}\underset{\sim}{b}}\, \overset{r}{a}_{\underset{\sim}{b}} , \tag{4.54}$$

where,

$$n r_{B}{}^{\underset{\sim}{a}\underset{\sim}{b}} = \iiiint \left(\overset{n}{g}{}''^{\,\underset{\sim}{a}} - \overset{n}{g}{}^{\underset{\sim}{a}} \right) f_o f_o' \, H_r{}^{\underset{\sim}{b}} W\pi^4 . \qquad (4.55)$$

The first few equations of the set (4.54) are,

$$N^a{}_{;a} = 0,$$

$$T^{ab}{}_{;b} = 0,$$

$$S^{abc}{}_{;c} = 2B^{abcd} a_{cd} . \qquad (4.56)$$

However from the condition $a_{abc}=0$, and equation (4.34') we can deduce that,

$$S^{abc} = S_o{}^{abc} + \alpha^{abc}{}_{de} \Delta T^{de} ,$$

and equation (4.56) can be written,

$$S_o{}^{abc}{}_{;c} + \left[\alpha^{abc}{}_{de} \Delta T^{de} \right]_{;c} = 2B^{abcd} a_{cd} , \qquad (4.57)$$

which is to be solved for ΔT^{de} in conjunction with,

$$M^{abcd} a_{cd} = \Delta T^{ab} . \qquad (4.33')$$

Thus we have 14 equations, $N^a{}_{;a}=0$, $T^{ab}{}_{;b}= 0$, and (4.57) for 14 unknowns, $u^a, n, T, \Delta T^{ab}$, the "relativistic 14-moment problem". *

It is clear that if the second term on the left hand side of equation (4.57) is neglected then the equation can be solved algebraically for ΔT^{ab} in terms of gradients of the equilibrium quantities, ie. we obtain a normal solution. Because B^{abcd} is essentially a collision frequency it is easy to see that the normal solution approximation is valid only when the microscopic time and length scales are much shorter than the macroscopic ones. In this case we obtain results equivalent to those derivable from the Chapman-Enskog theory. However our procedure has two important advantages:

---

*There is one more moment in the relativistic problem because of the extra variable $\pi := p^* - p$.

a) it is more general; f_0 need not be a local equilibrium distribution, but in the Chapman-Enskog expansion (4.3) the leading term must be a local equilibrium distribution, although the method used here has involved some complicated algebraic manipulation, the operations have been simple, and no integral equations have to be solved.

The Chapman-Enskog method has certain advantages:

a) the approximations involved in the Chapman-Enskog method are well understood, see eg. Grad (1963),

b) the Casimir-Onsager relations can be derived within the Chapman-Enskog theory, but a derivation based on the Grad method does not seem to be known.

If the normal solution approximation cannot be made then the first order partial differential equations must be solved. This approach will be discussed in section 4.7. The next section will be devoted to the usual, normal solution, transport equations for a 1-component gas.

4.6 The Transport Equations for a 1-Component Gas.

In this section we solve the equations,

$$S_0{}^{abc}{}_{;c} = 2B^{abcd}\, a_{cd}\,, \tag{4.58}$$

$$M^{abcd}\, a_{cd} = \Delta T^{ab}. \tag{4.33'}$$

which govern the behaviour of ΔT^{ab} in a normal solution.

If we assume that f_0 is a LDS distribution, its moments take the form,

$$mN_0{}^{a} = \rho u^{a}\,, \tag{4.59}$$

$$T_o^{ab} = (\mu+p)u^a u^b + pg^{ab}, \tag{4.60}$$

$$S_o^{abc}{}_{;} = \omega u^a u^b u^c + 3\xi u^{(a}g^{bc)} , \tag{4.61}$$

$$Q_o^{abcd} = \varphi u^a u^b u^c u^d + 6\chi g^{(ab}u^c u^{d)} + 3\psi g^{a(b}g^{cd)} . \tag{4.62}$$

The coefficients are not all independent, and by taking the various contractions we obtain,

$$p = \tfrac{1}{4}\left[(\mu+p)-A_o m^2\right], \tag{4.63}$$

$$\xi = 1/6\ (\omega - \rho m) , \tag{4.64}$$

$$\chi = 1/8\left[\varphi - (\mu+p)m^2\right] , \tag{4.65}$$

$$\psi = 1/48\left[\varphi - 3(\mu+p)m^2 + 2A_o m^4\right] . \tag{4.66}$$

For the remainder of this section we shall assume that f_o is a Maxwellian distribution of the form (2.37). Following Israel (1963) we first verify by direct integration that,

$$A_o = 4\pi m^2 e^{\alpha} K_1(\gamma)/\gamma , \tag{4.67}$$

where $\gamma := mc^2\beta = mc^2/kT$ is a dimensionless inverse temperature, and K_n is the modified Bessel function of order n defined by

$$K_n(z) = \frac{\pi^{\frac{1}{2}}(\frac{1}{2}z)^n}{\Gamma(n+\frac{1}{2})}\int_0^\infty e^{-z\cosh u}\sinh^{2n}u\,du .$$

By considering the integrals for $A_o, \rho, \mu+p, \omega, \varphi$ and differentiating (4.67) repeatedly with respect to γ or by direct integration, we also obtain,

$$\rho = 4\pi m^4 e^{\alpha} K_2(\gamma)/\gamma , \tag{4.68}$$

$$\mu+p = 4\pi m^4 e^{\alpha} K_3(\gamma)/\gamma , \tag{4.69}$$

$$\omega = 4\pi m^5 e^{\alpha} K_4(\gamma)/\gamma , \qquad (4.70)$$

$$\varphi = 4\pi m^6 e^{\alpha} K_5(\gamma)/\gamma , \qquad (4.71)$$

and from the recurrence formulae for the Bessel functions the other coefficients are,

$$p = 4\pi m^4 e^{\alpha} K_2(\gamma)/\gamma^2 , \qquad (4.72)$$

$$\xi = 4\pi m^5 e^{\alpha} K_3(\gamma)/\gamma^2 , \qquad (4.73)$$

$$\chi = 4\pi m^6 e^{\alpha} K_4(\gamma)/\gamma^2 , \qquad (4.74)$$

$$\psi = 4\pi m^6 e^{\alpha} K_3(\gamma)/\gamma^3 . \qquad (4.75)$$

These formulae are valid for $m \neq 0$ and in the limit as $m \to 0$. Since f_0 is a LDS distribution we also have the expansions,

$$\alpha^{ab}{}_c = C_1 u^a u^b u_c + C_2 g^{ab} u_c + C_3 u^{(a} \delta^{b)}{}_c , \qquad (4.76)$$

$$M^{abcd} = M_1 u^a u^b u^c u^d + M_2 g^{ab} u^c u^d + M_3 u^a u^b g^{cd} + M_4 u^{(a} g^{b)(c} u^{d)} + M_5 g^{a(c} g^{d)b} + M_6 g^{ab} g^{cd} , \qquad (4.77)$$

$$2B^{abcd} = B_1 u^a u^b u^c u^d + B_2 g^{ab} u^c u^d + B_3 u^a u^b g^{cd} + B_4 u^{(a} g^{b)(c} u^{d)} + B_5 g^{a(c} g^{d)b} + B_6 g^{ab} g^{cd} ,$$

where the C's and M's can be evaluated from the equations (4.23),(4.24), and the B's are certain collision integrals which will be evaluated in the next chapter. We expand ΔT^{ab} as,

$$\Delta T^{ab} = \tau(4u^a u^b + g^{ab}) + 2q^{(a} u^{b)} + \pi^{ab}, \qquad (4.78)$$

where τ is the difference in pressures between the actual and zero order distributions, and q^a, π^{ab} are the heat flux and stress tensor defined by equation (3.13). Differentiation of equation (4.61) gives,

$$S_o^{abc}{}_{;c} = (\dot{\omega} + \omega\theta)u^a u^b + (\dot{\xi} + \xi\theta)g^{ab} + 2\omega\dot{u}^{(a}u^{b)} + 2\xi_{,}^{(a}u^{b)} + 2\xi u^{(a;b)}, \quad (4.79)$$

where $\dot{\omega} := \omega_{,a}u^a$ etc., and the other quantities were defined in chapter 2.

We now make a further approximation (similar to neglecting ΔT^{ab} in (4.57)) and use the zeroth order conservation equations $T_o{}^{ab}{}_{;b} = 0$ to write $\dot{\xi}$ and the gradient of α in terms of $\theta, \dot{u}^a$ obtaining

$$S_o{}^{abc}{}_{;c} = X_1(4u^a u^b + g^{ab}) + 2X_2{}^{(a}u^{b)} + X_3{}^{ab}, \quad (4.80)$$

where,

$$X_1 \; := \; \xi\theta(5/3 - \Gamma - \Gamma/\gamma h), \quad (4.81)$$

$$X_2{}^a \; := \; -mph'\gamma^2 h^{ab}\left[\dot{u}_b/\gamma + (1/\gamma)_{,b}\right], \quad (4.82)$$

$$X_3{}^{ab} := 2\xi\sigma^{ab}, \quad (4.83)$$

and $h(\gamma) = (\mu + p)/(m\rho) = K_3(\gamma)/K_2(\gamma)$ is the <u>relativistic enthalpy</u>, ($= 1 +$ classical enthalpy $/c^2$), and Γ is the <u>ratio of specific heats</u>,

$$\Gamma := c_p/c_v = \gamma^2 h'/(\gamma^2 h' + 1),$$

where "'" = $d/d\gamma$. A detailed derivation is given in Israel (1963).

After some tedious algebra we obtain, using (4.77),(4.78),(4.80),

$$\tau = X_1(3M_2 + M_5)/(3B_2 + B_5), \quad (4.84)$$

$$q^a = X_2{}^a(2M_5 - M_4)/(2B_5 - B_4), \quad (4.85)$$

$$\pi^{ab} = X_3{}^{ab}M_5/B_5. \quad (4.86)$$

These three equations contain the usual transport equations in a slightly disguised form. For using (4.83) we have,

$$\pi^{ab} = -\eta(2\sigma^{ab}), \quad (4.87)$$

where η is the shear viscosity. The factor 2 is required in (4.87) so as to correspond to the classical definition, but otherwise this equation is identical to (3.36) and so provides a more rigorous justification for the phenomenological derivation given in chapter 3. Equations (4.82),(4.85) give,

$$q^a = -\lambda h^{ab}\left[\dot{u}_b/\gamma + (1/\gamma)_{,b}\right] , \qquad (4.88)$$

where,

$$\lambda := mph'\gamma^2(2M_5-M_4)/(2B_5-B_4), \qquad (4.89)$$

is the relativistic thermal conductivity. This equation is also idential to the phenomenological analogue developed in chapter 3. Finally from equations (4.81), (4.84) we obtain,

$$\tau = \theta\xi(5/3 - \Gamma - \Gamma/\gamma h)(3M_2+M_5)/(3B_2+B_5). \qquad (4.90)$$

In the equilibrium theory the pressure and internal energy per unit mass e are related by the equation of state,

$$p = p(\rho, e) \qquad (4.91)$$

For a general non-equilibrium state, there is no equation of state, but if e is known, equation (4.91) determines a pressure p^* (cf. equation (3.17)). This is not the same as the kinetic pressure p in general, although $p^*=p$ for a classical monatomic gas, and for a radiation gas. In analogy with the other transport equations we might expect the difference to be proportional to a scalar depending on the relative velocity field, ie. θ and so ζ is defined by,

$$\pi := p^* - p = \zeta\theta , \qquad (4.92)$$

(cf. equation (3.37)). ζ is called the bulk viscosity, (see eg. Lighthill, (1953)). In relativity theory the perfect gas law holds in equilibrium,

$p = nkT$ $\qquad$ $n=$ number density

as can be deduced from equations (4.68), (4.72). For a small change in T which keeps n constant,

$$\begin{aligned} \Delta p &= nk\,\Delta T , \\ &= \rho(\Gamma-1)c_v\Delta T, \text{ since } c_p - c_v = k/m,\ \rho = nm, \\ &= \rho(\Gamma-1)\Delta e, \text{ by definition of } c_v \end{aligned}$$

Thus,

$$\Delta p = (\Gamma - 1)\Delta\mu \quad \text{since} \quad \mu = \rho(1+e).$$

Since the energy density of our gas is $\mu_0 + 3\tau$ the thermodynamic pressure is $p^* = p_0 + 3\tau(\Gamma - 1)$. The kinetic pressure is $p = p_0 + \tau$, and so,

$$\pi = p^* - p = (3\Gamma - 4)\tau .$$

Therefore equation (4.92) holds in the relativistic case with,

$$\zeta = \xi(5/3 - \Gamma - \Gamma/\gamma h)(3\Gamma - 4)(3M_2 + M_5)/(3B_2 + B_5), \tag{4.93}$$

as the <u>relativistic bulk viscosity coefficient</u>.

Although the coefficient is not the same in different schemes, the total pressure (equilibrium + correction) has the same value when calculated in each of the schemes.

4.7 Discussion of the Transport Equations: Two Paradoxes Resolved.

As a summary of section 4.6 we repeat the transport equations,

$$\pi^{ab} = -2\eta\sigma^{ab} , \tag{4.94}$$

$$q^a = -\lambda\, h^{ab} \left[\dot{u}_b/\gamma + (1/\gamma)_{,b}\right] , \tag{4.95}$$

$$\pi = \zeta\theta . \tag{4.96}$$

The shear viscosity equation corresponds very closely to the classical one, and we shall not discuss it in this section.

The first paradox that can be resolved was mentioned as a theorem in chapter 2. A relativistic gas, whose particles have non-zero rest mass, which is expanding, cannot remain in equilibrium; that is, isentotropic expansion is impossible. The validity of this result has been questioned by Schücking & Spiegel (1970). However it is easy to see from equations (4.92),(4.93) that within the context of kinetic theory this effect is unavoidable because in general the bulk viscosity coefficient ς is non-zero. However in the ultra-relativistic limit $\Gamma \to 4/3$ and so $\varsigma \longrightarrow 0$, (see equation (4.93),) while in the classical limit, $\Gamma \longrightarrow 5/3$, $h \to 1$, $\gamma \longrightarrow \infty$ and so again $\varsigma \longrightarrow 0$, ie. isentotropic expansion* is possible only in these two limiting cases. Furthermore as Stewart et al (1970) have shown this result could be expected to hold in any physically reasonable more general statistical mechanics. (I thank Prof.Ehlers for suggesting the improved argument presented here). To see this we suppose that the 1-particle distribution function f can depend on other parameters besides the coordinates of the point in 8-dimensional phase space at which it is evaluated. We denote averaging over all of these other parameters and over momentum space with f as weight function by $\langle\ \rangle$. If $\underline{v}$ is the 3-velocity of an individual particle, and $\hat{\gamma} = (1-v^2/c^2)^{-\frac{1}{2}}$ it is easily verified that,

$$\mu/\rho = 1 + e = \langle \hat{\gamma} \rangle$$

$$3p/\rho = \langle \hat{\gamma} v^2 \rangle = \langle \hat{\gamma} - 1/\hat{\gamma} \rangle = 1 + e - \langle 1/\hat{\gamma} \rangle .$$

Now suppose f is varied in such a way that ρ and $\langle\hat{\gamma}\rangle = 1+e$ remain constant. Clearly $\langle 1/\hat{\gamma} \rangle$ will not remain constant in general and so p is not determined by ρ, e alone. In the ultrarelativistic limit $\hat{\gamma} \to \infty$, $3p/\rho \to 1+e$ and $p=p(\rho, e)$. In the classical limit $\hat{\gamma} \sim 1+v^2/c^2$, $\langle 1/\hat{\gamma} \rangle \approx 1$, and so $p = p(\rho, e)$. However between these limits ρ, e are not sufficient to determine p and so there is a bulk viscosity effect.

Although the heat conduction equation (4.95) contains a term proportional to the temperature gradient there is also a term proportional to the acceleration of the fluid which does not occur in the classical theory. However there is an implicit factor $1/c^2$ multiplying it. This term occurs because an

---

* (at a finite rate)

acceleration of matter can be considered as an energy flux which ought to be included in q^a. Eckart (1940) has described this term as being due to the inertia of energy. Another way of seeing the necessity of term has been suggested by Ehlers. If the simple Fourier law was valid then in the absence of heat conduction γ or T would be constant, $h^{ab}T_{,b}=0$. However this would contradict the well known Tolman temperature-redshift relation implicit in (2.41). Moreover the heat conduction equation leads to a paradox because just as in the classical theory we may show that q^a obeys a parabolic partial differential equation, and this would allow an infinite propagation velocity for heat signals. This has been thought to be a defect of relativistic thermodynamics since the propagation velocity should be finite,(see eg. Zumino (1957), Cattaneo (1958), Vernotte (1958), Hughes (1961), Kranys (1966,1967)). Cattaneo, Vernotte and Kranys have proposed that the equation (4.95) should be modified to read,

$$q^a + \kappa \dot{q}^a = -\lambda h^{ab} \left[\dot{u}_b/\gamma + (1/\gamma)_{,b}\right] . \qquad (4.97)$$

If this were done, and if κ were positive, then q^a would satisfy a hyperbolic equation and would have a finite propagation velocity. However there is no good reason for introducing such a term; it was postulated on a purely ad hoc basis.

If the parabolic equation were taken seriously, the large signal speeds would contradict the previously made assumption that the microscopic time and length scales were much shorter than the macroscopic ones. Consequently it would no longer be permissible to use normal solution theory, and instead the full equation (4.57) would have to be used. We now study this equation in more detail. Eliminating a_{cd} between equations (4.57) and (4.33') we can rewrite (4.57) as,

$$\alpha^{abcde} \Delta T_{de;c} + X^{abde} \Delta T_{de} = Y^{ab}, \qquad (4.98)$$

where Y^{ab} is independent of ΔT^{ab}. Equation (4.98) is a first order linear partial differential equation for ΔT_{de}. To discuss it we first make some inessential but simplifying assumptions. We consider a 1-dimensional wave, (in the x^1 direction,) in Minkowski space, so that $g_{ab}=\mathrm{diag}(-1,1,1,1)$. We adopt barycentric coordinates so that $u^a = (1,0,0,0)$. We also define

$$\varphi^{abde} := \alpha^{abcde}\,\varphi_{,c}$$

where φ is for the moment an arbitrary scalar field. The system (4.98) contains only five independent equations according to our assumptions, given by ab = 00,01,02,03,11, and we relabel these by A= 1,2,3,4,5 respectively. There are only five independent components of T^{de} given by, de = 00,01,02,03,11, and we relabel these by B= 1,2,3,4,5 respectively. Then according to standard theory (see eg. Courant & Hilbert (1962),) the characteristics of the system (4.98) are the surfaces φ = constant, where φ satisfies,

$$\det(\varphi^{AB}) = 0, \quad A,B=1,2,3,4,5. \tag{4.99}$$

To solve this equation we first expand α^{abcde} as,

$$\alpha^{abcde} = \left[D_1u^au^bu^c+3D_2g^{(ab}u^{c)}\right]u^du^e + \left[D_3u^au^bu^c+3D_4g^{(ab}u^{c)}\right]g^{de} +$$
$$+6\left[D_5u^{(a}u^b+D_6g^{(ab}\right]g^{c)(d}u^{e)}+6D_7u^{(a}g^{b(d}g^{e)c)} \,. \tag{4.100}$$

It is then easily found that the non-zero components of ^{AB} are,

(where $\dot{\varphi} := \varphi_{,0}$, $\varphi' := \varphi_{,1}$)

$$\varphi^{11} = (D_1-3D_2-D_3+3D_4-6D_5+6D_6+6D_7)\dot{\varphi} = a_1 \text{ say,}$$
$$\varphi^{12} = (D_5-D_6-2D_7)\,\varphi' = a_2,$$
$$\varphi^{15} = (D_3-3D_4)\,\dot{\varphi} = a_3,$$
$$\varphi^{21} = (D_2-D_4-2D_6)\,\varphi' = a_4,$$
$$\varphi^{22} = \varphi^{33} = \varphi^{44} = (D_5-D_6-2D_7)\dot{\varphi} = a_5,$$
$$\varphi^{25} = (D_4+2D_7)\,\varphi' = a_6 ,$$
$$\varphi^{51} = (D_2-D_4-2D_6)\dot{\varphi} = a_7,$$
$$\varphi^{52} = 3D_6\,\varphi' = a_8 ,$$
$$\varphi^{55} = (D_4+2D_7)\dot{\varphi} = a_9$$

Evaluation of the determinant gives,

$$\begin{vmatrix} a_1 & a_4 & 0 & 0 & a_7 \\ a_2 & a_5 & 0 & 0 & a_8 \\ 0 & 0 & a_5 & 0 & 0 \\ 0 & 0 & 0 & a_5 & 0 \\ a_3 & a_6 & 0 & 0 & a_9 \end{vmatrix} = 0,$$

which eventually reduces to,

$$\dot{\varphi} = 0, \ \underline{\text{or}} \quad 3D_6 \varphi'^2 - (D_5 - D_6 - 2D_7)\dot{\varphi}^2 = 0. \qquad (4.101)$$

Assuming for the moment that $(D_5 - 2D_7/D_6) > 1$ we see that equation (4.98) has real characteristic surfaces, and furthermore, the maximum velocity of propagation of finite amplitude non-adiabatic perturbations is,

$$v_{max} = (3D_6/(D_5 - D_6 - 2D_7))^{\frac{1}{2}} \ . \qquad (4.102)$$

The problem now is to show that $v_{max}^2 > 0$, and to evaluate v_{max}. We make use of the expansions (4.61), (4.62), (4.67) together with the formulae (4.73)-(4.75), and the expansion,

$$P_0^{abcde} = P_1 u^a u^b u^c u^d u^e + 10 P_2 u^{(a} u^b u^c g^{de)} + 15 P_3 u^{(a} g^{bc} g^{de)} \ ,$$

for the fifth moment of f_0 where,

$$P_2 = 4\pi m^7 e^{\alpha} K_5(\gamma)/\gamma^2 \ , \qquad P_3 = 4\pi m^7 e^{\alpha} K_4(\gamma)/\gamma^3 .$$

It is then easily verified that

$$D_5 = (P_2 - P_3 M_4/M_5 - \xi \chi/p)/(M_5 - \tfrac{1}{2} M_4),$$

$$D_6 = (P_3 - \xi \chi/p)/(M_5 - \tfrac{1}{2} M_4),$$

$$D_7 = P_3/M_5,$$

so that

$$\frac{3D_6}{D_5 - D_6 - 2D_7} = \frac{3(K_2 K_4 - K_3^2)}{\gamma(K_2 K_5 - K_3 K_4) - 3K_2 K_4 + K_3^2} = \frac{3}{5} \ . \qquad (4.103)$$

Thus we have proved,

Theorem

Although in the normal solution approximation the Grad theory gives a parabolic equation for heat conduction, the general theory gives a hyperbolic equation for the propagation of finite amplitude (but small) non-adiabatic perturbations of a Maxwellian gas from thermal equilibrium. The maximum velocity of propagation is given by equations (4.102),(4.103) and does not exceed,

$$v_{max} = c(3/5)^{\frac{1}{2}} \approx 0.8c .$$

4.8 Matching Conditions (continued).

We return here to the problem of matching conditions which was mentioned briefly in section 4.4. It will be remembered that the most convenient matching condition for our purpose was,

$$A_0 = A, \qquad N_0{}^a = N^a . \tag{4.35}$$

This will be called the Grad matching; it was also used by Marle (1969). If we write $n_G = n_0$ etc to describe the matching, equation (4.35) is equivalent to,

$$n_G = n, \qquad u_G{}^a = u^a , \qquad \mu_G - 3p_G = \mu - 3p. \tag{4.104}$$

An alternative scheme, based on the phenomenological theory of Eckart (1940), can be written,

$$n_E = n , \quad u_E{}^a = u^a , \qquad \mu_E = \mu \quad , \tag{4.105}$$

and a third scheme in which the definition of the mean velocity introduced by Landau & Lifschitz (1959) is used, can be written

$$\mu_L u_L{}^a = N^a - q_L{}^a , \quad \mu_L u_L{}^a = T^{ab} u_{Lb} \, . \tag{4.106}$$

If q^a, π, π^{ab} are regarded as small compared with the equilibrium quantities, and $\approx$ denotes equality to first order, then it is easy to see, (eg. Marle(1966),) that,

$$
\begin{aligned}
n_G &= n_E \approx n_L \\
u_G^a &= u_E^a \approx u_L^a + q_L^a/\rho_L \\
q_G{}^a &= q_E{}^a \approx q_L{}^a/h_L \\
\pi_G^{ab} &= \pi_E^{ab} \approx \pi_L^{ab} \\
\mu_G + 3\tau_G &= \mu_E \approx \mu_L , \\
p_G + \tau_G &= p_E + \tau_E = p_L + \tau_L , \\
-(3\Gamma_G - 4)\tau_G &\approx \tau_E \approx \tau_L
\end{aligned}
$$

Thus the definitions of n, A_0 used by Eckart and by Landau & Lifschitz agree with ours to first order, and so therefore does their definition of the temperature. It is then a trivial matter to verify that the transport laws for shear viscosity and heat flow derived by the other authors are the same as ours, with the same numerical coefficient (to zeroth order). We assume that since the gradients of the equilibrium quantities are small (first order), the gradients of the non-equilibrium, first order quantities are of second order smallness. Since the transport laws are of the form,

(non-equilibrium quantity) = (coefficient)x(gradient of macroscopic variable),

the coefficient is only required to zeroth order accuracy.

The bulk viscosity is somewhat more complicated, since p^* will take different values in the different schemes, and so π_G will differ from π_E, π_L by a factor different from 1. In both the Eckart and Landau & Lifschitz schemes, the zeroth order energy density μ_0 is μ, while in the Grad scheme, $\mu_0 = \mu - 3\tau$. Thus the argument between equations (4.92),(4.93) is not needed in their schemes, and so we have,

$$\pi = (3\Gamma_G - 4)\tau_G \approx \tau_E \approx \tau_L \; .$$

Thus the bulk viscosity equations are equivalent. This gives the justification for our earlier assertion that all three matching schemes were equivalent.

4.9 <u>The 1-Component Quantum Gas.</u>

We now show how to include quantum mechanical effects into the theory. For a comprehensive treatment of the non-relativistic theory see eg. Uehling & Uhlenbeck (1933,1934). For simplicity we only consider the 1-component gas. The Boltzmann equation is,

$$L(f) = \iiint (\hat{f}\hat{f}' f'' f''' - f f' \hat{f}'' \hat{f}''')W \pi' \pi'' \pi''' \; ,$$

where $\hat{f}' := 1 \pm f'$ etc., and the upper sign refers to bosons, the lower to fermions. By comparison with the non-quantum equations we see that quantum effects are unimportant if $f \ll 1$. The formulae (4.67)-(4.75) are inappropriate here, because they were calculated for a Maxwellian distribution. If instead we use a Bose-Einstein (Fermi-Dirac) distribution for

$$f_0(x,p) = [\exp(-\alpha - \beta_a p^a) \mp 1]^{-1} \; , \qquad (2.34)$$

then equations (4.67)-(4.71) are replaced by,

$$A_0 = I_0(\beta), \rho = I_1(\beta), \mu + p = I_2(\beta), \omega = I_3(\beta), \varphi = I_4(\beta), \qquad (4.107)$$

where,

$$I_n(\beta) := 4\pi \int \frac{E^n \pi}{\exp(\beta E - \alpha) \mp 1} \qquad (4.108)$$

and $E := -u_a p^a$, $\beta = 1/kT$. The other coefficients can be obtained from the relations (4.63)-(4.66). Just as in section (4.2) we multiply the Boltzmann equation by each $\overset{n}{g}{}^{\underset{\sim}{a}}$ in turn and integrate over the mass shell to obtain,

$$\int \overset{n}{g}{}^{\underset{\sim}{a}} L(f)\pi = \iiiint \overset{n}{g}{}^{\underset{\sim}{a}}(\hat{f}\,\hat{f}' f'' f''' - f\,f'\hat{f}''\hat{f}''')W\pi^4 .$$

Using the reversibility and symmetry of collisions we obtain,

$$\left[\int \overset{n}{g}{}^{\underset{\sim}{a}} p^c f\,\pi\right]_{;c} = \iiint (\overset{n}{g}{}''^{\underset{\sim}{a}} - \overset{n}{g}{}^{\underset{\sim}{a}}) f\,f'\hat{f}''\hat{f}'''W\pi^4 . \qquad (4.109)$$

The first few equations in the scheme (4.109) are,

$$N^a{}_{;a} = 0,$$

$$T^{ab}{}_{;b} = 0,$$

$$S^{abc}{}_{;c} = \iiiint (p''^a p''^b - p^a p^b) f\,f'\hat{f}''\hat{f}'''W\pi^4 . \qquad (4.110)$$

We now expand each of the four distribution functions occuring on the right hand side of equation (4.110) and linearise and truncate as before to obtain, (in the normal solution approximation,)

$$S_o{}^{abc}{}_{;c} = 2B^{*abcd}\,a_{cd} , \qquad (4.111)$$

where,

$$B^{*abcd} := B^{abcd} \pm C^{abcd},$$

and,

$$C^{abcd} := \iiiint (p''^a p''^b - p^a p^b) H^{cd} f_o f'_o f''_o f'''_o W\pi^4 . \qquad (4.112)$$

On comparing equation (4.11) with (4.58) we see that the discussion of transport processes follows as before. The introduction of quantum mechanical effects necessitates only two modifications. The functions entering into the numerators of the transport coefficients are given in terms of the $I_n(\beta)$

of equation (4.107) rather than by K Bessel functions, and the collision integrals require a correction term C^{abcd} .

4.10 The Transport Equations for a Multicomponent Gas.

Applying the normal solution approximation to the set of equations (4.40) gives

$$\Omega_{Ao} := \sum_{B,C,D} \left\{ {}^{000}C_{ABCD} - {}^{000}D_{ABCD} + {}^{000}E_{ABC} - {}^{000}F_{ABC} + {}^{00}G_{ABC} - {}^{00}K_{ABC} \right\},$$

$$\Omega_A = Ao + \sum_E \left[{}^{01}B^{c}_{AE}a_{Ec} + {}^{02}B^{cd}_{AE}a_{Ecd} \right], \qquad (4.113)$$

$$T^{ab}_{Ao;b} = \sum_E \left[{}^{11}B^{ac}_{AE}\, a_{Ec} + {}^{12}B^{acd}_{AE}\, a_{Ecd} \right], \qquad (4.11\ )$$

$$S^{abc}_{Ao\ ;c} = \sum_E \left[{}^{21}B^{abc}_{AE}\, a_{Ec} + {}^{22}B^{abcd}_{AE}\, a_{Ecd} \right], \qquad (4.11\ )$$

which are to be solved in conjunction with,

$$M^{ab}_A a_{Ab} = \varepsilon_A u^a + J^a_A \ , \qquad (4.32''')$$

$$M^{abcd}_A a_{Acd} = \Delta T^{ab}_A - \alpha^{ab}_{A\ c}(\varepsilon_A u^c + J^c_A). \qquad (4.33''')$$

The method of solution is very similar to that in the 1-component case. We first make the expansions,

$$M^{ab}_A = M_{A11}u^a u^b + M_{A12}g^{ab} , \qquad (4.116)$$

$$M^{abcd}_A = M_{A21}u^a u^b u^c u^d + M_{A22}g^{ab}u^c u^d + M_{A23}u^a u^b g^{cd} + M_{A24}u^{(a}g^{b)(c}u^{d)} +$$

$$+ M_{A25}g^{a(c}g^{d)b} + M_{A26}g^{ab}g^{cd} , \qquad (4.117)$$

$$\alpha^{ab}_{A\ c} = C_{A1}u^a u^b u_c + C_{A2}g^{ab}u_c + C_{A3}u^{(a}\delta^{b)}_{\ c} , \qquad (4.118)$$

where the coefficients can be obtained from the defining relations (4.23), (4.24) once each f_{Ao} has been given. We also write,

$$a_{Ac} = K_{A1}u_c + K_{A2c} \quad , \text{ with } \quad K_{A2c}u^c = 0, \tag{4.119}$$

$$\begin{aligned} a_{Acd} &= L_{A1}(4u_c u_d + g_{cd}) + 2L_{A2(c}u_{d)} + L_{A3cd} \text{ , with} \\ L_{A2c}u^c &= L_{A3cd}u^d = L_{A3c}{}^c = 0, \end{aligned} \tag{4.120}$$

$$\begin{aligned} \Delta T_A^{ab} &= \tau_A(4u^a u^b + g^{ab}) + 2q_A^{(a}u^{b)} + \pi_A^{ab} \text{ , with} \\ q_{Aa}u^a &= \pi_{Aab}u^b = \pi_{Aa}{}^a = 0. \end{aligned} \tag{4.121}$$

Here $\tau_A, q_A{}^a, \pi_A{}^{ab}$ are the pressure difference, heat flow and shearing stress tensor of the A-component of the gas relative to the mean flow of the gas. Equations (4.32'''), (4.33''') can now be solved to give,

$$K_{A1} = \varepsilon_A/(M_{A12} - M_{A11}), \tag{4.122}$$

$$K_{A2}^a = J_A^a/M_{A12}, \tag{4.123}$$

$$L_{A1} = \tau_A/(3M_{A22} + M_{A25}), \tag{4.124}$$

$$L_{A2}^a = (2q_A^a - C_{A3}J_A^a)/(2M_{A25} - M_{A24}), \tag{4.125}$$

$$L_{A3}^{ab} = \pi_A^{ab}/M_{A25} \;. \tag{4.126}$$

If we make the substitutions,

$$ {}^{01}B^{c}_{AE} = {}^{01}B_{AE}u^{c}, \tag{4.127} $$

$$ {}^{02}B^{cd}_{AE} = {}^{02}B_{AE1}u^{c}u^{d} + {}^{02}B_{AE2}g^{cd}, \tag{4.128} $$

$$ {}^{11}B^{cd}_{AE} = {}^{11}B_{AE1}u^{c}u^{d} + {}^{11}B_{AE2}g^{cd}, \tag{4.129} $$

$$ {}^{12}B^{acd}_{AE} = {}^{12}B_{AE1}u^{a}u^{c}u^{d} + (\tfrac{1}{4}{}^{12}B_{AE1} - \tfrac{1}{2}{}^{12}B_{AE2})u^{a}g^{cd} + 2\,{}^{12}B_{AE2}g^{a(c}u^{d)}, \tag{4.130} $$

$$ {}^{21}B^{abc}_{AE} = {}^{21}B_{AE1}u^{a}u^{b}u^{c} + (\tfrac{1}{4}{}^{21}B_{AE1} - \tfrac{1}{2}{}^{21}B_{AE2})g^{ab}u^{c} + 2\,{}^{21}B_{AE2}g^{c(a}u^{b)}, \qquad (4.130) $$

$$ \begin{aligned} {}^{22}B^{abcd}_{AE} = {} & {}^{22}B_{AE1}u^{a}u^{b}u^{c}u^{d} + {}^{22}B_{AE2}g^{ab}u^{c}u^{d} + {}^{22}B_{AE3}u^{a}u^{b}g^{cd} + \\ & + {}^{22}B_{AE4}u^{(a}g^{b)(c}u^{d)} + {}^{22}B_{AE5}g^{a(c}g^{d)b} + {}^{22}B_{AE6}g^{ab}g^{cd}. \end{aligned} \tag{4.131} $$

where the coefficients are integrals which have to be determined, and the expansions,

$$ T^{ab}_{Ao;b} = T_{A1}u^{a} + T^{a}_{A2}, \text{ where } T^{a}_{A2}u_{a} = 0, \tag{4.132} $$

$$ \begin{aligned} & S^{abc}_{Ao\,;c} = S_{A0}u^{a}u^{b} + 4S_{A1}g^{ab} + 2S^{(a}_{A2}u^{b)} + S^{ab}_{A3}, \\ & \quad \text{where, } \quad S^{a}_{A2}u_{a} = S^{ab}_{A3}u_{b} = S^{a}_{A3a} = 0, \end{aligned} \tag{4.133} $$

then we can expand equations (4.113)-(4.115) to obtain,

$$ \Omega_{A} = \Omega_{Ao} + \sum_{E}\Big[-{}^{01}B_{AE}K_{E1} + 12\,{}^{02}B_{AE}L_{E1}\Big], \tag{4.134} $$

$$ T_{A1} = \sum_{E}\Big[-{}^{11}B_{AE1}K_{E1} + 3({}^{12}B_{AE1} - 2\,{}^{12}B_{AE2})L_{E1}\Big], \tag{4.135} $$

$$ T^{a}_{A2} = \sum_{E}\Big[{}^{11}B_{AE2}K^{a}_{E2} + 2\,{}^{12}B_{AE2}L^{a}_{E2}\Big], \tag{4.136} $$

$$ S_{A1} = \sum_{E}\Big[-{}^{21}B_{AE2}K_{E1} + (3\,{}^{22}B_{AE2} + {}^{22}B_{AE5})L_{E1}\Big], \tag{4.137} $$

$$S^{a}_{A2} = \sum_{E}\left\{{}^{21}B_{AE3}K^{a}_{E2} + \tfrac{1}{2}(2\,{}^{22}B_{AE5} - {}^{22}B_{AE4})L^{a}_{E2}\right\}, \tag{4.138}$$

$$S^{ab}_{A3} = \sum_{E}\left\{{}^{22}B_{AE5}L^{ab}_{E3}\right\} . \tag{4.139}$$

We can eliminate the K's and L's from these equations by using the equations (4.122)-(4.126), and obtain,

$$\Omega_{A} = \Omega_{Ao} + \sum_{E}\left\{-\frac{{}^{01}B_{AE}\,\varepsilon_{E}}{M_{E12}-M_{E11}} + 12\,\frac{{}^{02}B_{AE}\,\tau_{E}}{3M_{E22}+M_{E25}}\right\} \tag{4.140}$$

$$T_{A1} = \sum_{E}\left\{-\frac{{}^{11}B_{AE1}\,\varepsilon_{E}}{M_{E12}-M_{E11}} + 3\,\frac{({}^{12}B_{AE1}-2\,{}^{12}B_{AE2})\,\tau_{E}}{3M_{E22}+M_{E25}}\right\}, \tag{4.141}$$

$$T^{a}_{A2} = \sum_{E}\left\{\left[\frac{{}^{11}B_{AE2}}{M_{E12}} - 2\,\frac{{}^{12}B_{AE3}C_{E3}}{2M_{E25}-M_{E24}}\right]J^{a} + 4\,\frac{{}^{12}B_{AE2}\,q^{a}_{E}}{2M_{E25}-M_{E24}}\right\}, \tag{4.142}$$

$$S_{A1} = \sum_{E}\left\{-\frac{{}^{21}B_{AE2}\,\varepsilon_{E}}{M_{E12}-M_{E11}} + \frac{(3\,{}^{22}B_{AE2}+{}^{22}B_{AE5})\,\tau_{E}}{3M_{E22}+M_{E25}}\right\}, \tag{4.143}$$

$$S^{a}_{A2} = \sum_{E}\left\{\left[\frac{{}^{21}B_{AE3}}{M_{E12}} - \frac{\tfrac{1}{2}(2\,{}^{22}B_{AE5}-{}^{22}B_{AE4})C_{E3}}{2M_{E25}-M_{E24}}\right]J^{a} + \frac{(2\,{}^{22}B_{AE5}-{}^{22}B_{AE4})\,q^{a}_{E}}{2M_{E25}-M_{E24}}\right\}, \tag{4.144}$$

$$S^{ab}_{A3} = \sum_{E}\left\{\frac{{}^{22}B_{AE5}\,\pi^{ab}_{E}}{M_{E25}}\right\} \tag{4.145}$$

We assume that each f_{Ao} is a local Boltzmann distribution so that,

$$T^{ab}_{Ao} = (\mu_{Ao}+p_{Ao})u^{a}u^{b} + p_{Ao}g^{ab} , \tag{4.146}$$

where the notation follows that of equations (4.59)-(4.62), and in the non-quantum case the coefficients are given by equations (4.68)-(4.75) if each γ is replaced by $\gamma_{A} := m_{A}c^{2}/kT$. Then it is easily verified that,

$$T^{ab}_{Ao\ ;b} = [\dot{\mu}_{Ao}+(\mu_{Ao}+p_{Ao})\theta]u^a+h^{ab}p_{Ao,b}+(\mu_{Ao}+p_{Ao})\dot{u}^a \ .$$

Thus,

$$T_{A1} = \dot{\mu}_{Ao}+(\mu_{Ao}+p_{Ao})\theta, \qquad (4.147)$$

$$T^a_{A2} = h^{ab}p_{Ao,b}+(\mu_{Ao}+p_{Ao})\dot{u}^a=-p_{Ao}h^{ab}(X_{Ab}+h_A X_{Ob}), \qquad (4.148)$$

where $X_{Ab}:=-\alpha_{A,b}$ and $X_{Ob} := \beta_{,b} - \beta\dot{u}_b$,($\beta = 1/kT$) were defined by equations (3.24), (3.25), and $h_A:=(\mu_{Ao}+p_{Ao})/\rho_{Ao}$ is the enthalpy of the A-component.

In a similar way we have,

$$S^{abc}_{Ao} = \omega_{Ao}u^a u^b u^c + 3\,\xi_{Ao}u^{(a}g^{bc)} \ ,$$

and,

$$S^{abc}_{Ao\ ;c} = S_{A1}(4u^a u^b+g^{ab}) +2S^{(a}_{A2}\,u^{b)} + S^{ab}_{A3} \ , \qquad (4.149)$$

where,

$$S_{A1} = \dot{\xi}_{Ao}+ 5/3\,\xi_{Ao}\theta \ , \qquad (4.150)$$

$$S^a_{A2} = h^{ab}\,\xi_{Ao}\left[X_{Ab} -(\omega_{Ao}/\xi_{Ao}-1)X_{Ob}\right], \qquad (4.151)$$

$$S^{ab}_{A3} = 2\,\xi_{Ao}\sigma^{ab} \ . \qquad (4.152)$$

We can now obtain the transport equations.

a) Shear Viscosity

In equation (4.145) we regard $({}^{22}B_{AE5}/M_{E25})$ as the (A,E) component of a NxN matrix. (N is the number of components in the gas.)

Assuming that this matrix is non-singular it has an inverse, π_{AE} and so equation (4.145) can be inverted to give,

$$\pi_A^{\ ab} = \sum_E \pi_{AE}\, s_{E3}^{\ ab}$$

$$= \sum_E \pi_{AE}\, \xi_{Eo}(2\sigma^{ab}) \,, \qquad \text{from equation (4.152),}$$

so that,

$$\pi^{ab} = \sum_A \pi_A^{\ ab} = \sum_A \sum_E \pi_{AE}\, \xi_{Eo}(2\sigma^{ab}),$$

and we have the transport equation,

$$\pi^{ab} = -\eta\,(2\sigma^{ab}), \tag{4.153}$$

where $\eta := -\sum_{A,E} \pi_{AE}\, \xi_{Eo}$ is the shear viscosity coefficient. Equation (4.153) provides the justification for the equation (3.36) obtained in chapter 3 from a thermodynamic argument.

b) Thermal Conductivity & Diffusion

Equations (4.142),(4.144) can be regarded as a set of 2N equations for the 2N unknown vectors $q_E^{\ a}, J_E^{\ a}$. Assuming the system is non-degenerate they will have a solution of the form,

$$q_A^{\ a} = \sum_E \left\{ q_{AE1} T_{E2}^{\ a} + q_{AE2} S_{E2}^{\ a} \right\},$$

$$J_A^{\ a} = \sum_E \left\{ J_{AE1} T_{E2}^{\ a} + J_{AE2} S_{E2}^{\ a} \right\},.$$

If we write $J_0^{\ a} := q^a := \sum_A q_A^{\ a}$ then it is easy to see using equations (4.148),(4.151) that we may write the solution as,

$$J_B^{\ a} = \sum_{C=0}^{N} L_{BC} h^{ab} X_{Cb} \,, \quad B=0,1,2,\ldots,N, \tag{4.154}$$

which should be compared with equation (3.39) derived earlier. It has not been shown here that L_{BC} is a symmetric matrix, which would imply the Casimir-Onsager relations, but such a result has not been found even in the classical treatment. Its verification might require a different approach to the problem.

c) Chemical Reactions & Bulk Viscosity

Before treating equations (4.141), (4.143) we must discuss the term

This term, occurs because according to the matching conditions (4.38') the gas mixture is not necessarily in chemical equilibrium although all components have equilibrium distributions with a common temperature.

From (4.113) we may obtain,

$$\Omega_{Ao} = \tfrac{1}{2}\sum_{B,C,D}\iiiint (f_{Co}f_{Do}-f_{Ao}f_{Bo})W\pi^4 + \tfrac{1}{2}\sum_{B,C}\iiint (f_{Bo}f_{Co}-f_{Ao})W\pi^3 - \sum_{B,C}\iiint (f_{Ao}f_{B_o}-f_{C_o})W\pi^3 \quad (4.155)$$

and using the matching conditions, we obtain,

$$\Omega_{Ao} = \tfrac{1}{2}\sum_{B,C,D}(e^{\alpha_C+\alpha_D}-e^{\alpha_A+\alpha_B})k_r(T) \ +2 \text{ similar terms} \quad (4.156)$$

where $k_r(T) := \iiiint \exp[-(E_A+E_B)/kT]\, W_r \pi^4$ is a function of temperature alone, depending on the type r of the reaction. If

$|\alpha_C+\alpha_D| \ll 1$, and $|\alpha_A+\alpha_B| \ll 1$,

$$e^{\alpha_C+\alpha_D} - e^{\alpha_A+\alpha_B} \approx \alpha_C+\alpha_D-\alpha_A-\alpha_B \approx \mathcal{A}_r/kT,$$

where $\mathcal{A}_r$ is the chemical affinity defined in chapter 3.

Thus

$$\Omega_{Ao} \approx \tfrac{1}{2}\sum_{B,C,D}\mathcal{A}_r k_r(T)/kT + 2 \text{ similar terms.} \quad (4.156')$$

If instead of summing over species in (4.157) we sum over a set of n independent reactions we obtain to first order in the affinities,

$$\Omega_{Ao} \approx \sum_{r=1}^{n} \nu_{Ar}\mathcal{A}_r k_r(T)/kT \quad (4.158)$$

where ν_{Ar} is the stoichiometric coefficient of the A-component in the rth reaction, $\mathcal{A}_r$ is the affinity of the rth reaction and k_r is a coefficient which depends only on temperature and densities, and so is independent of affinities of concentrations or time. From equation (3.27) we obtain

$$J_r = k_r \mathcal{A}_r / kT. \tag{4.158}$$

As is the classical case (see eg. de Groot & Mazur (1962)) the system of equations (4.158) gives

$$J_r = \sum_{s=1}^{n} \Lambda_{rs}^{(o)} \mathcal{A}_s / kT, \quad r=1,2,\ldots,n, \tag{4.159}$$

where the coefficients $\Lambda_{rs}^{(o)}$ are functions of T independent of the $\mathcal{A}_r$'s. This equation is identical to the phenomenological equation (3.39) in the absence of expansion. An equivalent expression has been obtained in the classical case by Ross & Mazur (1961). However, as Ross & Mazur remark, there are correction terms to the $\Lambda_{rs}^{(o)}$ in equation (4.159) because of perturbations to the Maxwellian distribution, and they calculated these in the classical case using the Chapman-Enskog method.

We can easily calculate the correction terms in the Grad theory using equations (4.141), (4.143). These 2N linear equations for the 2N unknowns ε_A, τ_A can be solved in terms of T_{E1}, S_{E1} or a linear combination of $\dot{\alpha}_E, \dot{\beta}, \theta$. To first order in $\mathcal{A}_r$ we have

$$N_{Ao\,;a}^{a} = \dot{n}_A + n_A\theta = \Omega_A = \sum_r \nu_{Ar} J_r = \sum_r \sum_s \nu_{Ar} \Lambda_{rs}^{(o)} \mathcal{A}_s / kT,$$

and these equations can be used to express $\dot{\alpha}_A$ as a linear combination of $\dot{\beta}, \theta, \mathcal{A}_r$. We can also use the approximate (zero order) equation $u_a T^{ab}{}_{;b} = 0$ to express $\dot{\beta}$ as a multiple of θ. We therefore eventually obtain ε_A, τ_A as linear combinations of $\mathcal{A}_r, \theta$, where the coefficients now depend on the chemical potentials (and therefore on the affinities).

By a calculation very similar to that used in section 4.6 we obtain

$$\Pi_A := p_A^* - p_A = (3\Gamma_A - 4)\tau_A + (p_{Ao}/\rho_{Ao})\varepsilon_A,$$

and on summing over A,

$$\pi = \varsigma\theta + \sum_r \Lambda_r \mathcal{A}_r / T . \tag{4.160}$$

Substituting these expressions for ε_A, τ_A into equation (4.140) we obtain (after eliminating the Ω_A as before,)

$$J_r^{(1)} = -\Lambda_r' \theta / T + \sum_{s=1}^{n} \Lambda^{(1)}_{rs} \mathcal{A}_s / T , \tag{4.161}$$

and, on combining with (4.159)

$$J_r = -\Lambda'_r \theta / T + \sum_s \Lambda_{rs} \mathcal{A}_s / T . \tag{4.162}$$

Equations (4.160), (4.162) are identical with equations (3.38) (3.39) which were derived from a phenomenological argument. However as Ross & Mazur point out, the coefficients depend on affinity, and implicitly on time. The simple form (4.159) is only valid at the lowest order of approximation.

CHAPTER 5. THE EVALUATION OF THE COLLISION INTEGRALS

5.1 Introduction

The purpose of this chapter is to show how, through a convenient choice of variables of integration, the evaluation of the collision integrals used in the last chapter can be reduced to a manageable task. In particular if, as we shall assume, the relativistic scattering cross-section for each type of collision depends only on the magnitude of the relative momentum before the collision, and the scattering angle, each collision integral can be reduced to a double integral over these two variables. Once the relevant cross-sections have been prescribed, the integrals can be evaluated numerically or analytically.

5.2 The Choice of the Variables of Integration

We shall first consider a collision, (possibly inelastic,) in which incoming particles with rest masses m_A, m_B and 4-momenta $p_A{}^a, p_B{}^a$ interact to produce outgoing particles with rest masses m_C, m_D and 4-momenta $p_C{}^a, p_D{}^a$. Annihilation and creation processes will be considered later. We define the total momentum before and after the collision to be,

$$\bar{p}^a := p_A{}^a + p_B{}^a \quad ,$$

$$\bar{p}'^a := p_C{}^a + p_D{}^a \quad ,$$

and the relative momentum before and after the collision to be

$$g^a := p_A{}^a - p_B{}^a \quad ,$$

$$g'^a := p_C{}^a - p_D{}^a \quad .$$

Conservation of momentum implies,

$$\bar{p}^a = \bar{p}'^a \quad . \tag{5.1}$$

From the definitions it follows that,

$$\bar{p}^2 := -\bar{p}_a\bar{p}^a = 2(m_A{}^2+m_B{}^2) + g^2, \tag{5.2}$$

where $g^2:=g_a g^a$. We shall use the total and relative 4-momenta before and after the collision as new variables of integration. Since they satisfy the relations (5.2) and $\bar{p}^a g_a = m_B{}^2 - m_A{}^2$ we may choose 6 independent variables and we do so as follows. Because of (5.2) we may take $\bar{p}^\alpha$ as independent variables. Fixing a direction, (but not magnitude) for $\bar{p}^a$ we see that g^α lies on the fixed hypersurface, $\bar{p}^a g_a = m_B{}^2 - m_A{}^2$. On this hypersurface g^a is fixed by its magnitude g and angular coordinates θ^*, φ^* with respect to some fixed coordinate system. The six independent variables are therefore taken to be $(\bar{p}^1, \bar{p}^2, \bar{p}^3, g, \theta^*, \varphi^*)$. We can calculate the Jacobian directly but it is preferable to use the following short cut. We choose coordinates so that, (assuming $m_A, m_B \neq 0$,)

$$p_A{}^a = (m_A, 0, 0, 0),$$

$$p_B{}^a = (m_B \cosh y, m_B \underline{e} \sinh y),$$ $\underline{e}$ a unit spacelike vector.

Then

$$g^a = (m_A - m_B \cosh y, -m_B \underline{e} \sinh y),$$

and,

$$g^2 = 2m_A m_B \cosh y - (m_A{}^2 + m_B{}^2).$$

Now,

$$\begin{aligned} \pi_B &= d^3 p_B / |p_{B0}|, \\ &= m_B{}^3 \sinh^2 y \cosh y \, dy \, d\Omega / m_B \cosh y, \\ &= F_{AB}(g) \, g \, dg \, d\Omega / 2m_A{}^2, \end{aligned}$$

where $d\Omega$ is an area element on the unit sphere in the 3-space orthogonal to $p_A{}^a$, and,

$$F_{AB}(g) := \left[g^4 + 2g^2(m_A{}^2+m_B{}^2) + (m_A{}^2-m_B{}^2)^2\right]^{\frac{1}{2}},$$

$$= \left[\bar{p}^4 - 2\bar{p}^2(m_A{}^2+m_B{}^2) + (m_A{}^2-m_B{}^2)^2\right]^{\frac{1}{2}} . \qquad (5.3)$$

We now write $p_A{}^a = m_A \hat{p}_A{}^a$ where $\hat{p}_{Aa}\hat{p}_A{}^a = -1$. Then $\pi_A = m_A{}^2 \hat{\pi}$ where $\hat{\pi}$ is an area element on the unit mass shell $\hat{p}_{Aa}\hat{p}_A{}^a = -1$. Therefore,

$$\pi_A \pi_B = \tfrac{1}{2} F_{AB}\, g dg d\Omega \hat{\pi} .$$

Now it is evident geometrically that

$$\pi_A \pi_B = A(g) dg d\Omega^* \bar{\pi} ,$$

where A (g) is to be determined, $d\Omega^* := \sin\Theta^* d\Theta^* d\varphi^*$ and, (for fixed g) $\bar{\pi}$ is an area element on the mass shell,

$$\bar{p}_a\bar{p}^a = -Q^2(g) := -2(m_A{}^2+m_B{}^2) - g^2.$$

If we define q^a by $\bar{p}^a = Qq^a$ then $\bar{\pi} = Q^2 \pi_q$ where π_q is an area element on the unit mass shell $q_a q^a = -1$. Thus,

$$\pi_A \pi_B = A(g) Q^2(g)\; dg d\Omega^* \pi_q .$$

To evaluate $A(g)$ we consider

$$I := \iint F(g)\, \pi_A \pi_B = \tfrac{1}{2} \iint F_{AB}(g) F(g) g dg d\Omega \hat{\pi} ,$$

$$= \iint F(g) Q^2(g) A(g) dg d\Omega^* \pi_q ,$$

for arbitrary integrable functions $F(g)$. Clearly the $d\Omega, d\Omega^*$ integrations are independent of g and can be taken over the same range. The same holds true for the $\hat{\pi}$, π_q integrations. Therefore,

$$A(g) = gF_{AB}(g)/2Q^2(g),$$

and

$$\pi_A \pi_B = \tfrac{1}{2} F_{AB}\, g\, dg\, d\Omega\, \bar{\pi} / Q^2, \tag{5.4}$$

$$= \tfrac{1}{2} F_{AB}\, g\, dg\, d\Omega\, \pi_q . \tag{5.4'}$$

Although the form (5.4') is preferable, because the π_q, dg integrations can be done in any order, we shall use the form (5.4) which is more convenient. In this case the integral over the mass shell $\bar{p}^a \bar{p}_a = -Q^2(g)$ must be done holding g constant ie. before the dg integration.

Equation (5.4) is considerably simplified if $m_A = m_B = m$, since then,

$$\pi_A \pi_B = \tfrac{1}{2} g^2\, dg\, d\Omega\, \bar{\pi} / \bar{p} \quad . \tag{5.5}$$

This form was first obtained by Israel (1963). Neither (5.4) nor (5.5) is strictly valid if m_A or m_B is zero. But it is easy to verify that the correct formulae are given by the appropriate limits:

$$m_B = 0,\ m_A \neq 0, \qquad \pi_A \pi_B = \tfrac{1}{2} g\, dg\, d\Omega\, \bar{\pi} \quad , \tag{5.4''}$$

$$m_A = m_B = 0, \qquad \pi_A \pi_B = \tfrac{1}{2} g\, dg\, d\Omega\, \bar{\pi} \quad . \tag{5.5'}$$

5.3 The Scattering Cross Section

It is now necessary to relate the scattering probability density $W_{AB\to CD}$ introduced in chapter 2 to the more usual scattering cross-section $\sigma_{AB\to CD}$. The differential cross-section $dQ(u)$ for collisions of the specified kind with respect to an observer having 4-velocity u^a is defined by,

$$dQ(u) := \frac{\text{spacetime density of collisions}}{\text{(density of target particles)x(relative flux density of incident particles)}}, \tag{5.6}$$

assuming that the final states are unpopulated, (see eg. Møller (1945), Terrall (1970), Ehlers (1971).) The observed spatial density of A-particles with momenta in the range π_A is,

$$n_A = f_A(x,p_A)(-u_a p_A{}^a)\pi_A ,$$

and similarly,

$$n_B = f_B(x,p_B)(-u_a p_B{}^a)\pi_B .$$

The relative velocity of the two particles, as measured by the observer is,

$$|v_B - v_A| = \frac{|(u_a p_A{}^a)p_B{}^b - (u_a p_B{}^a)p_A{}^b|}{(u_a p_A{}^a)(u_a p_B{}^a)}$$

a formula which can easily be verified in the rest frame of the observer. Using (5.6) and the definition of $W_{AB\to CD}$, (2.18), we obtain,

$$dQ(u)\,|(u_a p_A{}^a)p_B{}^b - (u_a p_B{}^a)p_A{}^b| = W_{AB\to CD}\pi_C\pi_D . \qquad (5.7)$$

It is easy to see that if u^a is a linear combination of $p_A{}^a, p_B{}^a$ then $dQ(u)$ is independent of u^a Thus the differential cross-section takes the same value dQ when measured in the rest frame of either particle or in the centre of mass frame. This value dQ is usually called the <u>relativistic cross-section.</u> (See e.g. Møller (1945), Terrall (1970).) Since it is a differential 2- form, it is often more convenient to derive a scalar from it, and to exhibit the $\delta^4(\bar{p}' - \bar{p})$ factor in $W_{AB\to CD}$ more explicitly. Accordingly we define the <u>relativistic scalar cross-section</u> by,

$$\sigma_{AB\to CD}\,\delta^4(\bar{p}^a - \bar{p}'^a)\,\pi_C\pi_D = dQ_{AB\to CD}. \qquad (5.8)$$

It follows from (5.7),(5.8) that,

$$W_{AB\to CD} = \tfrac{1}{2}F_{AB}\,\delta^4(\bar{p}^a - \bar{p}'^a)\,\sigma_{AB\to CD} . \qquad (5.9)$$

In the special case $m_A = m_B$ we have,

$$W_{AB\to CD} = \tfrac{1}{2} gp\, \delta^4(\bar{p}^a - \bar{p}'^a)\, \sigma_{AB\to CD}\,, \tag{5.10}$$

so that our definition of the scalar cross-section agrees with that of Bichteler (1965) for the equal mass case. It is an easy matter to check that in the classical limit $\sigma_{AB\to CD}$ reduces to the usual classical value.

Finally we write,

$$\delta^4(\bar{p}^a - \bar{p}'^a)\, \pi_C \pi_D = \delta_{\bar{\pi}}(\bar{p}^\alpha - \bar{p}'^\alpha)\, \delta(g' - [g^2 + 2(m_A^2 + m_B^2 - m_C^2 - m_D^2)]^{\frac{1}{2}})\, \bar{\pi}'\, dg'\, d\Omega',$$

where $\delta_{\bar{\pi}}(\bar{p}^\alpha - \bar{p}'^\alpha)$ is a distribution on the mass shell $\bar{p}^a \bar{p}_a = -Q^2$ defined by,

$$\int G(p'^a)\, \delta_{\bar{\pi}}(\bar{p}^\alpha - \bar{p}'^\alpha)\, \bar{\pi}' = G(\bar{p}^a). \qquad \text{for arbitrary } G.$$

Then,

$$W_{AB\to CD}\, \pi_A \pi_B \pi_C \pi_D = \tfrac{1}{4} F_{AB}^2\, g\, \sigma_{AB\to CD}\, \delta_{\bar{\pi}}(\bar{p}^\alpha - \bar{p}'^\alpha)\, \times$$

$$\times\, \delta(g' - [g^2 + 2(m_A^2 + m_B^2 - m_C^2 - m_D^2)]^{\frac{1}{2}})\; Q^{-2}\, \bar{\pi}'\, dg'\, d\Omega'\, \bar{\pi}\, dg\, d\Omega\,. \tag{5.11}$$

It should be noted that the $d\Omega$ integration is taken over a unit sphere in the 3-space $\bar{p}^a g_a = m_B^2 - m_A^2$. Therefore if

$$g^a = G_{AB}\bar{p}^a + H_{AB} n^a, \tag{5.12}$$

where n^a is a spacelike unit vector orthogonal to $\bar{p}^a$ and,

$$G_{AB} := (m_A^2 - m_B^2)/\bar{p}^2,$$

$$H_{AB} := F_{AB}/\bar{p}\,,$$

then the $d\Omega$ integration can be regarded as an integration over the unit sphere spanned by n^a.

5.4 The Integration Procedure: 1-Component Case.

For the sake of lucidity we first consider the 1-component gas, although the principle is the same in the multicomponent case. There is one collision integral,

$$I := {}^{22}B^{abcd} := \tfrac{1}{2}\iiiint (p^{*a}p^{*b} + p^{**a}p^{**b} - p^{a}p^{b} - p'^{a}p'^{b}) H^{cd} f_o f_o' W\pi^4 \;. \qquad (4.55)$$

As in chapter 4 we shall assume for simplicity that,

$$f_o = \exp(\alpha + \beta_a p^a), \qquad (5.13)$$

so that,

$$f_o f_o' = \exp(2\alpha + \beta_a \bar{p}^a). \qquad (5.14)$$

From formulae (5.2),(5.3) we have, assuming the particles have rest mass m

$$\bar{p} = (g^2+4m^2)^{\frac{1}{2}},$$

$$F_{AB} = g(g^2+4m^2)^{\frac{1}{2}}.$$

Consequently, $\quad G_{AB} = 0, \quad H_{AB} = g,$

and the splitting (5.12) of g^a gives,

$$g^a = gn^a.$$

This simplicity occurs because $\bar{p}^a, g^a$ are orthogonal. Formula (5.11) gives,

$$W\pi^4 = \tfrac{1}{4}g^3\sigma\,\delta_{\bar{\pi}}(\bar{p}^\alpha - \bar{p}'^\alpha)\,\delta(g'-g)\,\bar{\pi}'dg'd\Omega'\,\bar{\pi}\,dgd\Omega.$$

Thus the integral can be written,

$$I = 1/32\; e^{2\alpha}\int^{(12)}_{\cdots\cdots}\!\!\int\Big\{\bar{p}'^a\bar{p}'^b + 2g'\bar{p}'^{(a}n'^{b)} + g'^2n'^an'^b + \bar{p}'^a\bar{p}'^b - 2g'\bar{p}'^{(a}n'^{b)} +$$

$$+ g'^2n'^an'^b - \bar{p}^a\bar{p}^b - 2g\bar{p}^{(a}n^{b)} - g^2n^an^b - \bar{p}^a\bar{p}^b + 2g\bar{p}^{(a}n^{b)} + g^2n^an^b\Big\}\Big\{\tfrac{1}{4}\big[\bar{p}^c\bar{p}^d +$$

$$+ 2g\bar{p}^{(c}n^{d)} + g^2n^cn^d\big] - \tfrac{1}{2}\alpha^{cd}{}_e(\bar{p}^e + gn^e - 2\alpha^e) - \beta^{cd}\Big\}\exp(\beta_a\bar{p}^a)g^3\sigma\,\delta_{\bar{\pi}}(\bar{p}'^\alpha - \bar{p}^\alpha)\times$$

$$\times\,\delta(g'-g)\,\bar{\pi}'dg'd\Omega'\,\bar{\pi}dgd\Omega.$$

(5.15)

The $\overline{\pi}'$, dg' integrations are trivial and give,

$$I = 1/16\, e^{2\alpha} \int^{(8)}_{\cdots\cdots} \int (n'^a n'^b - n^a n^b) \left\{ \tfrac{1}{4} \left[\bar{p}^c \bar{p}^d + 2g\bar{p}^{(c} n^{d)} + g^2 n^c n^d \right] - \right.$$

$$\left. -\tfrac{1}{2}\, \alpha^{cd}{}_e (\bar{p}^e + gn^e - 2\alpha^e) - \beta^{cd} \right\} \exp(\beta_a \bar{p}^a) g^5 \sigma \, d\Omega' \, d\Omega \, \overline{\pi}\, dg. \qquad (5.16)$$

We now write $d\Omega' = \sin\Theta \, d\Theta \, d\Phi$ where Θ, Φ are the polar angles of n'^a with respect to n^a so that Θ is the scattering angle. As stated earlier we are assuming $\sigma = \sigma(g, \Theta)$ ie. the collisions are axisymmetric. Now the Φ integration can be accomplished giving,

$$\int_0^{2\pi} d\Phi\, (n'^a n'^b - n^a n^b) = \sin^2\Theta (\bar{h}^{ab} - 3n^a n^b),$$

where $\bar{h} := \bar{p}^a \bar{p}^b / \bar{p}^2 + g^{ab}$ is the projection operator into the spacelike hypersurface orthogonal to $\bar{p}^a$. Then,

$$I = 1/16\, \pi e^{2\alpha} \int^{(7)}_{\cdots\cdots} \int (\bar{h}^{ab} - 3n^a n^b) \left\{ \tfrac{1}{4} \left[\bar{p}^c \bar{p}^d + 2g\bar{p}^{(c} n^{d)} + g^2 n^c n^d \right] - \tfrac{1}{2} \alpha^{cd}{}_e (\bar{p}^e + gn^e - \right.$$

$$\left. -2\alpha^e) - \beta^{cd} \right\} \exp(\beta_a \bar{p}^a) g^5 \sigma \sin^3\Theta \, d\Theta \, \overline{\pi} \, dg\, d\Omega. \qquad (5.17)$$

The Ω integration is now straightforward since,

$$\int d\Omega = 4\pi,$$

$$\int d\Omega\, n^a = 0,$$

$$\int d\Omega\, n^a n^b = 4\pi/3\, \bar{h}^{ab},$$

$$\int d\Omega\, n^a n^b n^c = 0,$$

$$\int d\Omega\, n^a n^b n^c n^d = 4\pi/5\, \bar{h}^{a(b} h^{cd)},$$

and gives,

$$I = \pi^2/240\, e^{2\alpha} \int^{(5)}_{\cdots\cdots} \int (5\bar{h}^{ab} h^{cd} - 9\bar{h}^{a(b} h^{cd)}) \exp(\beta_a \bar{p}^a) g^7 \times$$

$$\times \sigma \sin^3\Theta \, d\Theta \, dg \, \overline{\pi} \quad . \qquad (5.18)$$

We can write $\exp(\alpha + \beta_a \bar{p}^a) = \bar{f}_0$ where $\bar{f}_0$ is a locally Maxwellian distribution and relativistic inverse temperature $\gamma := (g^2 + 4m^2)^{\frac{1}{2}} c^2 / kT$. Denoting the moments of this distribution by, $\bar{A}_0$, $\bar{N}_0{}^a$, etc., we can easily do the π integration to give,

$$I = \pi^2/120 \;\; e^{\alpha} \iint \Big[-2\bar{Q}_0^{abcd}(g^2+4m^2)^{-2} + (\bar{T}_0^{ab}g^{cd} + g^{ab}\bar{T}_0^{cd})(g^2+4m^2)^{-1} - \\ -6\bar{T}_0^{(a(c}{}_g{}^{d)b)}(g^2+4m^2)^{-1} + (g^{ab}g^{cd} - 3g^{a(c}{}_g{}^{d)b})\bar{A}_0 \Big] g^7 dg \sin^3\Theta \, d\Theta \tag{5.19}$$

In chapter 4 we used the splitting,

$$2^{22}B^{abcd} = B_1 u^a u^b u^c u^d + B_2 g^{ab} u^c u^d + B_3 u^a u^b g^{cd} + B_4 u^{(a}{}_g{}^{b)(c} u^{d)} + \\ + B_5 g^{a(c}{}_g{}^{d)b} + B_6 g^{ab} g^{cd} .$$

It is now easily verified that,

$$\begin{aligned} B_1 &= -2(I_1+12I_2+48I_3), \\ B_2 &= I_1+2I_2-12I_3, \\ B_3 &= B_2, \\ B_4 &= -2(3I_1+16I_2+24I_3), \\ B_5 &= -(3I_1+6I_2+4I_3), \\ B_6 &= I_1+2I_2-2I_3, \end{aligned} \tag{5.20}$$

where,

$$I_n := m^2/15 \; \pi^3 e^{2\alpha} \int_0^{\infty} \int_0^{\pi} K_n(\gamma\tilde{Q})(\gamma\tilde{Q})^{-n} g^7 \tilde{Q}^2 \sigma(g,\Theta) \sin^3\Theta \, d\Theta \, dg, \tag{5.21}$$

with $\tilde{Q} := (4+g^2/m^2)^{\frac{1}{2}}$, $\gamma = mc^2/kT$.

These integrals have been evaluated for a particular value of the cross-section in the next section. It should be noted however that in spite of the superficial complexity it was comparatively easy to reduce the scattering integral to a manageable form because of the relation (5.14).

5.5 The Transport Coefficients for a 1-Component Gas with a Maxwellian Scattering Cross-section.

The simplest possible cross-section which could serve as an illustration is one independent of g This was originally mentioned by Maxwell (1867) who noted that $\sigma(g,\Theta)$ = constant corresponded to a r^{-5} potential. Although this is physically unrealistic it is a convenient example to take. However in generalising any classical cross-section to the relativistic case there is a certain amount of arbitrariness. For example we may include arbitrary factors of $\frac{1}{2}Q$ in the relativistic generalisation, since $\frac{1}{2}Q \rightarrow 1$ in the classical limit. We therefore consider a cross-section independent of g defined by,

$$\sigma_0 = 1/(2\pi)\int_0^{2\pi} \sigma(\Theta)\sin^3\Theta \, d\Theta = \text{constant}. \quad (5.22)$$

Then the formula (5.21) reduces to,

$$I_n = 1/30\,\pi^2 m^2 e^{2\alpha}\sigma_0 \int_{2\gamma}^{\infty} K_n(\gamma\tilde{Q})(\gamma\tilde{Q})^{-n}\tilde{Q}^2 g^7 dg \,. \quad (5.23)$$

Writing $x = \gamma\tilde{Q}$, equation (5.23) simplifies to give,

$$I_n = \frac{2^5 m^2 \pi^2 e^{2\alpha}}{15\,\gamma^4}\sigma_0 \int_{2\gamma}^{\infty} K_n(x)x^{3-n}\left[\frac{x^2}{4\gamma^2} - 1\right]^3 dx. \quad (5.24)$$

The integral is a standard one, and using equations (9.6.26), (11.3.28) of Abramowitz & Stegun (1965) we can obtain,

$$I_n = \frac{2^7 m^2 \pi^2 e^{2\alpha}}{5\times 2^n \gamma^{n+4}}\sigma_0\left[K_{4-n}(2\gamma) + 4K_{5-n}(2\gamma)/\gamma\right]. \quad (5.25)$$

It is now possible to write out the transport coefficients as explicit functions of γ. There seems to be little point in doing this in the general case, but the classical and ultrarelativistic limits are of interest. In the classical case we have:

i) shear viscosity,

i) shear viscosity,

$$\eta = \frac{5m}{24\sigma_0}\left(\frac{\pi}{\gamma}\right)^{\frac{1}{2}}\left\{1 + \frac{25}{16}\frac{1}{\gamma} - \frac{293}{1536}\frac{1}{\gamma^2} - \frac{20159}{24576}\frac{1}{\gamma^3} + \cdots\right\}, \quad (5.26)$$

ii) thermal conductivity,

$$\lambda = \frac{25m}{32\sigma_0}\left(\frac{\pi}{\gamma}\right)^{\frac{1}{2}}\left\{1 + \frac{13}{16}\frac{1}{\gamma} - \frac{1951}{512}\frac{1}{\gamma^2} + \frac{57335}{8192}\frac{1}{\gamma^3} + \cdots\right\}, \quad (5.27)$$

iii) bulk viscosity,

$$\zeta = \frac{25m}{96\sigma_0}\left(\frac{\pi}{\gamma^5}\right)^{\frac{1}{2}}\left\{1 - \frac{183}{16}\frac{1}{\gamma} + \frac{41001}{512}\frac{1}{\gamma^2} - \frac{3635165}{8192}\frac{1}{\gamma^3} + \cdots\right\}. \quad (5.28)$$

These series are all asymptotic rather than convergent. The leading terms in the shear viscosity agree with the classical theory, (see eg. Grad (1949)) and $\zeta/\eta = O(\gamma^{-2})$ vanishes in the classical limit. This verifies that the theory developed in this study is correct at least in the classical limit.

In the ultrarelativistic limit we have:

i) shear viscosity,

$$\eta = 16m/(5\sigma_0)\left(1 + \frac{1}{20}\gamma^2 + \frac{1}{50}\gamma^4\log\gamma + O(\gamma^4)\right), \quad (5.29)$$

ii) thermal conductivity,

$$\lambda = 16m/\sigma_0\left(1 - \frac{3}{4}\gamma^2 - \gamma^4\log\gamma + O(\gamma^4)\right), \quad (5.30)$$

iii) bulk viscosity,

$$\zeta = m\gamma^3/(9\sigma_0)\left(1 + 6\gamma^2\log\gamma + O(\gamma^2)\right). \quad (5.31)$$

In this case η tends to infinity like T, λ tends to a constant, and ζ tends to zero like T^{-3} as $T \rightarrow \infty$. However it should not be supposed that this asymptotic behaviour would occur in general. These formulae were calculated for a mathematically convenient but physically unrealistic cross-section.

5.6 The Integration Procedure: Multicomponent Case.

In the study of a multicomponent gas many collision integrals occur. We shall not evaluate all of them, but merely consider two examples which demonstrate the techniques used to deal with all of them,

$$\text{a)}\quad {}^{220}C^{abcd}_{ABCD} = \tfrac{1}{2}\iiiint g^{2ab}_{A}\, H^{cd}_{C}\, f_{Co} f_{Do} W_{AB\to CD}\, \pi^{4} \quad , \tag{4.41}$$

$$\text{b)}\quad {}^{22}G^{abcd}_{ABC} = \iiint g^{ab}_{A} H^{cd}_{C} f_{Co} W_{AB\to C}\, \pi^{3} \; . \tag{4.45}$$

a) The Evaluation of ${}^{220}C^{abcd}_{ABCD}$.

As in section (5.4) we shall assume that the distribution functions occurring in these collision integrals are locally Maxwellian, so that,

$$f_{Co} f_{Do} = \exp(\alpha_C + \alpha_D + \beta_a \bar{p}'^{a}). \tag{5.32}$$

Using formulae (5.11), (5.12) and the $\bar{p}^{a}, g^{a}$ variables we have,

$$\begin{aligned} I := {}^{220}C^{abcd}_{ABCD} = \frac{1}{32}\exp(\alpha_C + \alpha_D) \int^{(12)}\!\!\cdots\!\!\int & \Big[\bar{p}^{a}\bar{p}^{b}(1+G_{AB})^{2} + 2(1+G_{AB})H_{AB}\bar{p}^{(a}n^{b)} + \\ & + H^{2}_{AB}n^{a}n^{b}\Big]\Big\{\tfrac{1}{4}\big[(1+G_{CD})^{2}\bar{p}'^{c}\bar{p}'^{d} + 2(1+G_{CD})H_{CD}\bar{p}'^{(c}n'^{d)} + H^{2}_{CD}n'^{c}n'^{d}\big] - \\ & - \tfrac{1}{2}\alpha^{cd}_{C\;e}\big[(1+G_{CD})\bar{p}'^{e} + H_{CD}n'^{e} - 2\alpha^{e}_{C}\big] - \beta^{cd}_{C}\Big\}\exp(\beta_a \bar{p}'^{a})\, g\bar{p}^{-2}F^{2}_{AB}\; \times \\ & \times \sigma_{ABCD}\, \delta_{\bar{\pi}}(\bar{p}'^{\alpha} - \bar{p}^{\alpha})\,\delta\big(g' - [g^{2} + 2(m^{2}_{A} + m^{2}_{B} - m^{2}_{C} - m^{2}_{D})]^{\frac{1}{2}}\big)\, \pi'\, dg'\, d\Omega'\, \bar{\pi}\, dg\, d\Omega . \end{aligned} \tag{5.33}$$

The $\bar{\pi}', dg'$ integrations are trivial and give,

$$\begin{aligned} I = \frac{1}{32}\exp(\alpha_C + \alpha_D) \int^{(8)}\!\!\cdots\!\!\int & \Big[(1+G_{AB})^{2}\bar{p}^{a}\bar{p}^{b} + 2(1+G_{AB})H_{AB}\bar{p}^{(a}n^{b)} + H^{2}_{AB}n^{a}n^{b}\Big] \times \\ & \times \Big\{\tfrac{1}{4}(1+G_{CD})^{2}\bar{p}^{c}\bar{p}^{d} + 2(1+G_{CD})H_{CD}\bar{p}^{(c}n'^{d)} + H^{2}_{CD}n'^{c}n'^{d} - \tfrac{1}{2}\alpha^{cd}_{C\;e}\big[(1+G_{CD})\bar{p}^{e} + \\ & + H_{CD}n'^{e} - 2\alpha^{e}_{C}\big] - \beta^{cd}_{C}\Big\} \exp(\beta_a \bar{p}^{a}) F^{2}_{AB}\, g\bar{p}^{-2}\sigma_{ABCD}\, d\Omega'\, \bar{\pi}\, dg\, d\Omega , \end{aligned} \tag{5.34}$$

where G_{CD}, H_{CD} are now functions of g. As before we write,

$$d\Omega' = \sin\Theta \, d\Theta \, d\Phi ,$$

where Θ is the scattering angle, and we assume $\sigma = \sigma(g, \Theta)$. Using the results,

$$\int_0^{2\pi} d\Phi \, n'^a = 2\pi \cos\Theta \, n^a ,$$

$$\int_0^{2\pi} d\Phi \, n'^a n'^b = \pi \sin^2\Theta (\bar{h}^{ab} - 3n^a n^b) + 2\pi n^a n^b ,$$

where $\bar{h}^{ab} := \bar{p}^a \bar{p}^b / \bar{p}^2 + g^{ab}$ as before, we obtain,

$$I = \frac{1}{32} \pi \exp(\alpha_C + \alpha_D) \int^{(7)} \cdots \int \Big[(1+G_{AB})^2 \bar{p}^a \bar{p}^b + 2(1+G_{AB}) H_{AB} \bar{p}^{(a} n^{b)} + H_{AB}^2 n^a n^b \Big] \times$$

$$\times \Big\{ \tfrac{1}{4} \Big[2(1+G_{CD})^2 \bar{p}^c \bar{p}^d + 4(1+G_{CD}) H_{CD} \cos\Theta \, \bar{p}^{(c} n^{d)} + H_{CD}^2 \sin^2\Theta (\bar{h}^{cd} - 3n^c n^d) +$$

$$+ 2H_{CD}^2 n^c n^d \Big] - \alpha_C{}^{cd}{}_e \Big[(1+G_{CD}) \bar{p}^e + H_{CD} \cos\Theta \, n^e - 2\alpha_C^e \Big] - 2\beta_C^{cd} \Big\} \exp(\beta_a \bar{p}^a) \times$$

$$\times F_{AB}^2 g \bar{p}^{-2} \sigma_{ABCD} \, d\Omega \, \bar{\pi} \, dg \sin\Theta \, d\Theta .$$

(5.35)

The $d\Omega$ integration gives

$$I = \frac{\pi^2}{480} \exp(\alpha_C + \alpha_D) \int^{(5)} \cdots \int \Big\{ (1+G_{AB})^2 \bar{p}^a \bar{p}^b \Big[\frac{15}{2} (1+G_{CD})^2 \bar{p}^c \bar{p}^d + \frac{5}{2} H_{CD}^2 \bar{h}^{cd} -$$

$$- 15 \alpha_C{}^{cd}{}_e \Big[(1+G_{CD}) \bar{p}^e - 2\alpha_C^e \Big] - 30 \beta_C^{cd} \Big] + 10(1+G_{AB}) H_{AB} (1+G_{CD}) H_{CD} \cos\Theta \, \bar{p}^{(a} \bar{h}^{b)(c} \bar{p}^{d)} -$$

$$- 10(1+G_{AB}) H_{AB} H_{CD} \cos\Theta \, \alpha_C{}^{cd}{}_e \bar{h}^{e(a} \bar{p}^{b)} + H_{AB}^2 \Big[\frac{5}{2} (1+G_{CD})^2 \bar{h}^{ab} \bar{p}^c \bar{p}^d +$$

$$+ \tfrac{1}{4} H_{CD}^2 \sin^2\Theta (5\bar{h}^{ab} \bar{h}^{cd} - 9\bar{h}^{a(b} \bar{h}^{cd)}) + \frac{3}{2} H_{CD}^2 \bar{h}^{a(b} \bar{h}^{cd)} - 5 \alpha_C{}^{cd}{}_e \Big[(1+G_{CD}) \bar{p}^e - 2\alpha_C^e \Big] \bar{h}^{ab} -$$

$$- 10 \bar{h}^{ab} \beta_C^{cd} \Big] \Big\} \exp(\beta_a \bar{p}^a) F_{AB}^2 g \bar{p}^{-2} \sigma_{ABCD} \, \bar{\pi} \, dg \sin\Theta \, d\Theta .$$

(5.36)

We may regard $\bar{f}_o := \exp(\beta_a \bar{p}^a)$ as a locally Maxwellian distribution with zero chemical potential and relativistic temperature $\bar{\gamma} := \bar{p}c^2/kT$. Denoting the moments of $\bar{f}_o$ by $\bar{A}_o, \bar{N}_o{}^a$ etc. the $\bar{\pi}$ integration gives

$$
\begin{aligned}
I = {} & \frac{\pi^2}{480}\exp(\alpha_C+\alpha_D)\iint\Big\{\bar{Q}_o^{abcd}\Big[\frac{15}{2}(1+G_{AB})^2(1+G_{CD})^2+\frac{5}{2}\bar{p}^{-2}\big\{(1+G_{AB})^2H_{CD}^2+(1+G_{CD})^2H_{AB}^2\big\}+\\
& +10\bar{p}^{-2}(1+G_{AB})H_{AB}(1+G_{CD})H_{CD}\cos\Theta+\bar{p}^{-4}H_{AB}^2H_{CD}^2\Big(\frac{3}{2}-\sin^2\Theta\Big)\Big]-5\bar{S}_o^{abc}\alpha_C{}^{cd}{}_e\times\\
& \times\Big[3(1+G_{AB})^2(1+G_{CD})+2\bar{p}^{-2}(1+G_{AB})H_{AB}H_{CD}\cos\Theta+\bar{p}^{-2}(1+G_{CD})H_{AB}^2\Big]+\bar{T}_o^{ab}g^{cd}\times\\
& \times\Big[\frac{5}{2}(1+G_{AB})^2H_{CD}^2+\frac{1}{2}\bar{p}^{-2}H_{AB}^2H_{CD}^2(1+\sin^2\Theta)\Big]+g^{ab}\bar{T}_o^{cd}\Big[\frac{5}{2}(1+G_{CD})^2H_{AB}^2+\frac{1}{2}\bar{p}^{-2}H_{AB}^2H_{CD}^2(1+\sin^2\Theta)\Big]\\
& +10\bar{T}_o^{ab}(\alpha_C{}^{cd}{}_e\alpha_C{}^e-\beta_C{}^{cd})\Big[3(1+G_{AB})^2+\bar{p}^{-2}H_{AB}^2\Big]+\bar{T}_o^{(a(c}{}_g{}^{d)b)}\times\\
& \times\Big[10(1+G_{AB})(1+G_{CD})H_{AB}H_{CD}\cos\Theta+3H_{AB}^2H_{CD}^2\cos^2\Theta\,\bar{p}^{-2}-H_{AB}^2H_{CD}^2\bar{p}^{-2}\Big]-\\
& -10\bar{N}_o^{(a}{}_g{}^{b)e}\alpha_C{}^{cd}{}_eH_{AB}H_{CD}\cos\Theta-5g^{ab}\alpha_C{}^{cd}{}_e\bar{N}_o^eH_{AB}^2(1+G_{CD})+\frac{3}{2}g^{a(b}{}_g{}^{cd)}\bar{A}_oH_{AB}^2H_{CD}^2\times\\
& \times\Big(1-\frac{3}{2}\sin^2\Theta\Big)+10H_{AB}^2g^{ab}(\alpha_C{}^{cd}{}_e\alpha_C{}^e-\beta_C{}^{cd})\bar{A}_o\Big\}F_{AB}^2g\bar{p}^{-2}\sigma_{ABCD}\,dg\sin\Theta\, d\Theta.
\end{aligned}
$$

(5.37)

Although this expression looks very complicated it can be decomposed in the same manner as before, and is only an integral over 2 variables, which can be performed either analytically or numerically once the cross-section has been specified. Furthermore in most practical cases considerable simplifications can be made. For example consider the reaction,

$$e^+ + e^- \rightarrow \gamma + \bar{\gamma}$$

Then $m_A=m_B=m$, $m_C=m_D=0$, $\alpha_C=\alpha_D=0$, and (5.37) simplifies to give,

$$
\begin{aligned}
I = {} & \frac{\pi^2}{480}\iint\Big\{\bar{Q}_o^{abcd}\Big(10(1+g/\bar{p}\cos\Theta)+g^2/\bar{p}^2\Big(\frac{3}{2}-\sin^2\Theta\Big)\Big)-5\bar{S}_o^{abe}\alpha_C{}^{cd}{}_e(3+2g/\bar{p}\cos\Theta+g^2/\bar{p}^2)\\
& +\frac{1}{2}\bar{T}_o^{ab}g^{cd}(5\bar{p}^2+g^2(1+\sin^2\Theta))+\frac{1}{2}g^{ab}\bar{T}_o^{cd}(5g^2+\bar{p}^2(1+\sin^2\Theta))+\\
& +10\bar{T}_o^{ab}(\alpha_C{}^{cd}{}_e\alpha_C{}^e-\beta_C{}^{cd})(3+g^2/\bar{p}^2)+\bar{T}_o^{(a(c}{}_g{}^{d)b)}(10g\bar{p}\cos\Theta+g^2(3\cos^2\Theta-1))-\\
& -10\bar{N}_o^{(a}{}_g{}^{b)e}\alpha_C{}^{cd}{}_eg\bar{p}\cos\Theta-5g^{ab}\alpha_C{}^{cd}{}_e\bar{N}_o^eg^2+\frac{3}{2}g^{a(b}{}_g{}^{cd)}\bar{A}_og^2\bar{p}^{-2}\Big(1-\frac{3}{2}\sin^2\Theta\Big)+\\
& +10g^2g^{ab}(\alpha_C{}^{cd}{}_e\alpha_C{}^e-\beta_C{}^{cd})\bar{A}_oF_{AB}^2g\bar{p}^{-2}\Big\}\sigma_{ABCD}\,dg\sin\Theta\, d\Theta.
\end{aligned}
$$

b) The Evaluation of ${}^{22}G^{abcd}_{ABC}$.

While we may replace the incoming momenta $p_A{}^a, p_B{}^a$ by $\bar{p}^a, g^a$ as before, the outgoing momentum requires slightly different treatment. The conservation of momentum is expressed by,

$$\bar{p}^a = p_C{}^a = \bar{p}'^a ,$$

which implies

$$g = \left[m_C^2 - 2(m_A^2 + m_B^2)\right]^{\frac{1}{2}} = g^* \quad \text{say.}$$

Then it is easy to see that,

$$W_{AB\to C}\pi^3 = \tfrac{1}{4}F^2_{AB}g\,\sigma_{ABC}\,\delta_{\bar{\pi}}(\bar{p}'^\alpha - \bar{p}^\alpha)\,\delta(g-g^*)Q^{-2}dg\,d\Omega\,\bar{\pi}\,\bar{\pi}' . \quad (5.38)$$

Assuming f_{Co} is locally Maxwellian we have,

$$I := {}^{22}G^{abcd}_{ABC} = \frac{1}{16}e^{\alpha_C}\int\overset{(9)}{\cdots\cdots}\int\left[(1+G_{AB})^2\bar{p}^a\bar{p}^b + 2(1+G_{AB})H_{AB}\bar{p}^{(a}n^{b)} + H^2_{AB}n^a n^b\right]\times$$
$$\times\left[\bar{p}'^c\bar{p}'^d - \alpha_C{}^{cd}{}_e(\bar{p}'^e - \alpha^e_C) - \beta^{cd}_C\right]F^2_{AB}gQ^{-2}\sigma_{ABC}\,\delta_{\bar{\pi}}(\bar{p}'^\alpha - \bar{p}^\alpha)\,\delta(g-g^*)\,\bar{\pi}'\,\bar{\pi}\,dg\,d\Omega .$$
(5.39)

The $\bar{\pi}'$, dg integrations are trivial and give,

$$I = \frac{1}{16}F^2_{AB}g^*\bar{p}^{-2}\exp(\alpha_C)\int\overset{(5)}{\cdots\cdots}\int\left[(1+G_{AB})^2\bar{p}^a\bar{p}^b + 2(1+G_{AB})H_{AB}\bar{p}^{(a}n^{b)} + \right.$$
$$\left. + H^2_{AB}n^a n^b\right]\left[\bar{p}^c\bar{p}^d - \alpha_C{}^{cd}{}_e(\bar{p}^e - \alpha^e_C) - \beta^{cd}_C\right]\sigma_{ABC}\,d\Omega\,\bar{\pi} . \quad (5.40)$$

The Ω integration is now straightforward and gives,

$$I = \tfrac{1}{4}\pi F^2_{AB}g^*\bar{p}^{-2}e^{\alpha_C}\iint\left[(1+G_{AB})^2\bar{p}^a\bar{p}^b + \tfrac{1}{3}H^2_{AB}\bar{h}^{ab}\right]\left[\bar{p}^c\bar{p}^d - \alpha_C{}^{cd}{}_e(\bar{p}^e - \alpha^e_C) - \beta^{cd}_C\right]\sigma_{ABC}\bar{\pi} .$$
(5.41)

where $\bar{h}^{ab} = \bar{p}^a\bar{p}^b/\bar{p}^2 + g^{ab}$ as before. Now the $\bar{\pi}$ integration gives,

$$I = \tfrac{1}{4}\pi F^2_{AB}g^*\bar{p}^{-2}\sigma_{ABC}\exp(\alpha_C)\left\{\left[(1+G_{AB})^2 + \tfrac{1}{3}H^2_{AB}/\bar{p}^2\right]\left[\bar{Q}_o^{abcd} - \bar{S}_o^{abe}\alpha_C{}^{cd}{}_e + \right.\right.$$
$$\left.\left. + \bar{T}_o^{ab}(\alpha_C{}^{cd}{}_e\alpha^e_C - \beta^{cd}_C)\right] + \tfrac{1}{3}H^2_{AB}g^{ab}\left[\bar{T}_o^{cd} - \alpha_C{}^{cd}{}_e\bar{N}_o^e + (\alpha_C{}^{cd}{}_e\alpha^e_C - \beta^{cd}_C)\bar{A}_o\right]\right\} . \quad (5.42)$$

It is obvious that for these annihilation and creation processes the cross-sections are constants.

We shall not discuss the general multicomponent case further. Unless there are certain simplifications, eg. $m_A = m_B$ the expressions clearly become unmanageable.

BIBLIOGRAPHY

Abramowitz M & Stegun I A, (1965) Handbook of Mathematical Functions (Dover).

Anderson J L, (1970) in Proceedings of the Midwest Conference on Relativity, (ed. M Carmeli et. al.)(Plenum Press).

Arzelies H,(1965) Nuovo Cim.,35,792.

Bakamjian B, Thomas L H,(1953) Phys.Rev., 92,1300.

Balescu R, Kotera T, (1967) Physica, 33,558.

Berezdivin R,& Sachs R K,(1970) in Proceedings of the Midwest Conference on Relativity, (ed. M Carmeli et.al.)(Plenum Press)

Bergmann P, (1951) Phys. Rev., 84,1026.

Bhatnagar D, Gross E, Krook M,(1954) Phys. Rev., 94,511.

Bichteler K, (1965) Ph.D. Dissertation, University of Hamburg, (unpublished).
(1965) Z. Physik 182,521
(1965) Comm. Math. Phys., 4,352.

Boyer R H, (1965) Amer. J. Phys., 33,910.

Burgers J M, (1969) Flow Equations for Composite Gases (Academic Press).

Carter B, (1969) Comm. Math. Phys., 10,280.

Cattaneo C, (1958) Compt.Rend., 247,431.

Chapman S, (1916) Phil.Trans.Roy.Soc., A216,279.

Chapman S, Cowling T G, (1970) Mathematical Theory of non-Uniform Gases (Cambridge).

Chernikov N A, (1963) Acta Physica Polonica,23,629.
Physics Letters 5,115.
(1964) Acta Physica Polonica 26,1069.
27,465.

Courant R, Hilbert D,(1962) Methods of Mathematical Physics II (Interscience).

Currie D G, (1963) J. Math.Phys.,4,1470.

Dam H Van, Wigner E P,(1965) Rev.Mod.Phys., 37,595.

Dantzig D Van,(1939) Proc.Kon.Ned.Akad.Wetensch., 42,608.

Dirac P A M,(1950) Canad. J. Math., 2,129.
(1951) Canad. J. Math., 3,1.

Eckart C, (1940) Phys.Rev., 58,919.

Ehlers J, (1961) Ahbandl,Akad.Wiss.Mainz,Math.Naturwiss,Kl.Nr.11.
(1971) in Proceedings Varenna Summer School on Relativistic Astrophysics 1969 (Academic Press).

Ehlers J, Geren P, Sachs R K, (1968) J.Math.Phys., 9,1344.

Enskog D, (1917) Dissertation, Uppsala University.

Fackerell E D, (1966) Ph.D. Dissertation,University of Sydney (unpublished).
(1968) Ap.J., 153,643.

Fackerell E D, Ipser J R, Thorne K S, (1969) Comments on Astrophysics & Space Science, 1,134.

Foldy L L, (1961) Phys.Rev., 122,275.

Gottal P de, Prigogine I, (1965) Physica, 31,677.

Grad H, (1949) Comm P. Appl.Math., 2, 325,331.
(1958) in Handbuch der Physik XII,(ed S Flugge) (Springer).
(1963) Physics of Fluids, 6,147.

Groot S R, Mazur P, (1962). Non-Equilibrium Thermodynamics, (N.Holland).

Gudehus T, Dubin D A, (1969) Preprint.

Hagedorn R,(1970) Astron. & Astrophys. 5,184.

Hakim R, (1967) J. Math.Phys., 8,1315,1379.
(1968) J. Math.Phys., 9,1805.

Havas P, (1965) in Statistical Mechanics of Equilibrium & Non-Equilibrium, (ed.J. Meixner) (N. Holland).

Hawking S W, (1968) Proc. Roy. Soc., A308,433.

Hilbert D, (1912) Math.Ann., 72,562.

Hughes W F, (1961) Proc.Cam.Phil.Soc., 57,878.

Israel W, (1963) J.Math.Phys., 4,1163.

Kadanoff L P, Baym G, (1962) Quantum Statistical Mechanics(Benjamin).

Kantowski R,(1965) Ph.D.Dissertation, University of Texas (unpublished).

Klimontovich Y L, (1958) Soviet Physics JETP, 34,119.
(1960) Soviet Physics JETP, 37,524.

Kluitenberg G , Groot S R de, (1954),Physica,20,199.

Kranys M, (1966) Nuovo Cim., B42,51.
(1967) Nuovo Cim., B50,48.

Krizan J E, (1965) Phys.Rev., 140,A1155.

Landau L D, Lifschitz E M, (1959) Fluid Mechanics (Pergamon).

Landsberg P T, Johns K A, (1970) Ann.Phys., 56,299.

Lichnerowicz A, Marrot R, (1940) Compt. Rend. 210,759.

Lighthill M J, (1953) in Surveys in Mechanics (ed. G K Batchelor) (Cambridge).

Lindquist R W, (1966) Ann.Phys., 37,487.

MacCallum M A H, (1970) Comm.Math.Phys. 19,31.

Mangeney A, (1964) Ph.D. Dissertation, Paris. See also, Ann.Phys.(Paris),10(1965).
(1970) Lectures, University of Cambridge.

Marle C, (1966) Compt. Rend. 263,485.
(1969) Ann.Inst. H. Poincare, 10,67,127.

Maxwell J C, (1867) Phil.Trans., 156,49.

Meixner J, (1958) Handbuch der Physik (ed. S. Flugge)(Springer)

Mintzner D, (1965) Phys. of Fluids, 8,1076.

Misner C W, (1968) Ap.J., 151,431.

Misner C W, Wheeler J A, Thorne K S,(1968) Open Letter to Relativity Theorists.

Møller C, (1945) Kgl.Dan.Vid. Selsk. 23,1.
(1967) Mat.Fys.Medd.Dan.Vid.Selsk., 36,nr 1.

Müller P, (1970) Diplomarbeit, University of Hamburg (unpublished).

Omnes R, (1970) On the origin of Matter & Galaxies (preprint).

Penrose O,(1970) Foundations of Statistical Mechanics, (Pergamon).

Prigogine I, (1965) in Statistical Mechanics of Equilibrium & Non-Equilibrium. (ed.J. Meixner) (N. Holland).

Robinson B B, Bernstein I B, (1962) Ann. Phys., 18,110.

Ross J, Mazur P, (1961) J. Chem. Phys., 35,19.

Sachs R K, (1971) in Proceedings Brandeis Summer School in Theoretical Physics 1968, Vol. 2.

Sasaki M, (1958) in Max Planck Festscrift (ed. W Frack) (DVW Berlin).

Schucking E L, Spiegel E A, (1970) Comments on Astrophysics & Space Science.1,121.

Stewart J M, (1969) Ph.D. Dissertation,University of Cambridge (unpublished).

Stewart J M, (1969) MNRAS 145,347.

Stewart J M, MacCallum M A H, Sciama D W, (1970) Comments on Astrophysics & Space Science, 2,206

Stewart J M, Anderson J L, (1971) paper in preparation.

Stueckelberg E C G, Wanders G,(1953) Helv.Phys.Acta, 26,307.

Synge J L, (1934) Trans.Roy.Soc.Canada, III, 28,127.

(1957) The Relativistic Gas (N. Holland).

(1964) Relativity - The General Theory (N.Holland).

Tauber G E, Weinberg J W, (1961) Phys. Rev., 122,1342.

Terrall J R, (1970) Amer J.Phys., 38,1460.

Thorne K S, (1967) in High Energy Astrophysics III (ed.B de Witt et al) (Gordon & Breach).

(1971) in Proceedings of Varenna Summer School on Relativistic Astrophysics 1969.

Tolman R C, (1934) Relativity, Thermodynamics & Cosmology (Oxford).

Treciokas R, Ellis G F R, (1971) preprint, University of Cambridge.

Uehling E A, Uhlenbeck G E, (1933) Phys.Rev. 43,552.

(1934) " " 46,917.

Vernotte P, (1958) Compt. Rend., 246,3154.

Walker A G, (1936) Proc.Edinburgh Math. Soc., 4,238.

Walker M, & Penrose T, (1970), Comm.Math.Phys., 18,265.

Zel'dovich Y B, Podurets M A, (1965) Soviet Astronomy AJ, 9,742.

Zumino B, (1957) Phys.Rev., 108,1116.

Beschaffenheit der Manuskripte

Die Manuskripte werden photomechanisch vervielfältigt; sie müssen daher in sauberer Schreibmaschinenschrift mit ausreichend großer Type geschrieben sein. Handschriftliche Formeln bitte nur mit schwarzer Tusche eintragen. Notwendige Korrekturen sind bei dem bereits geschriebenen Text entweder durch Überkleben des alten Textes vorzunehmen oder aber müssen die zu korrigierenden Stellen mit weißem Korrekturlack abgedeckt werden. Die reproduktionsfähigen Abbildungen (in Originalgröße) sollen in den Text eingeklebt werden. Falls das Manuskript oder Teile desselben neu geschrieben werden müssen, ist der Verlag bereit, dem Autor bei Erscheinen seines Bandes einen angemessenen Betrag zu zahlen. Die Autoren erhalten 50 Freiexemplare.

Zur Erreichung eines möglichst optimalen Reproduktionsergebnisses ist es erwünscht, daß bei der vorgesehenen Verkleinerung der Manuskripte der Text auf einer Seite in der Breite möglichst 18 cm und in der Höhe 26,5 cm nicht überschreitet. Entsprechende Satzspiegelvordrucke werden vom Verlag gern auf Anforderung zur Verfügung gestellt.

Manuskripte, in englischer, deutscher oder französischer Sprache abgefaßt, sind einzureichen bei: Springer-Verlag, 6900 Heidelberg, Postfach 1780.

Cette série a pour but de donner des informations rapides, de niveau élevé, sur des développements récents en physique, aussi bien dans la recherche que dans l'enseignement supérieur. On prévoit de publier.

1. des versions préliminaires de travaux originaux et de monographies
2. des cours spéciaux portant sur un domaine nouveau ou sur des aspects nouveaux de domaines classiques
3. des rapports de séminaires
4. des conférences faites lors de congrès ou de colloques

En outre il est prévu de publier dans cette série, si la demande le justifie, des rapports de séminaires et des cours multicopiés ailleurs mais déjà épuisés.

Dans l'intérêt d'une diffusion rapide, les contributions auront souvent un caractère provisoire; le cas échéant, les démonstrations ne seront données que dans les grandes lignes. Les travaux présentés pourront également paraître ailleurs. Une réserve suffisante d'exemplaires sera toujours disponible. En permettant aux personnes intéressées d'être informées plus rapidement, les éditeurs Springer espèrent, par cette série de «prépublications», rendre d'appréciables services aux instituts de physique. Les annonces dans les revues spécialisées, les inscriptions aux catalogues et les copyrights rendront plus facile aux bibliothèques la tâche de réunir une documentation complète.

Présentation des manuscrits

Les manuscrits, étant reproduits par procédé photomécanique, doivent être soigneusement dactylographiés type assez grand. Il est recommandé d'écrire à l'encre de Chine noire les formules non dactylographiées. Les corrections nécessaires doivent être effectuées soit par collage du nouveau texte sur l'ancien soit en recouvrant les endroits à corriger par du vernis correcteur blanc. Les illustrations; en dimension originale, préparées pour reproduction sont à insérer dans le texte. S'il s'avère nécessaire d'écrire de nouveau le manuscrit, soit complètement, soit en partie, la maison d'édition se déclare prête à verser à l'auteur, lors de la parution du volume, le montant des frais correspondants. Les auteurs recoivent 50 exemplaires gratuits.

Pour obtenir une reproduction optimale il est désirable que le texte dactylographié sur une page ne dépasse pas 26,5 cm en hauteur et 18 cm en largeur. Sur demande la maison d'edition met à la disposition des auteurs du papier spécialement preparé.

Les manuscrits en anglais, allemand ou français peuvent être adressés à Springer-Verlag, 6900 Heidelberg, Postfach 1780.